THE RISE AND IMPENDING DEMISE OF

ISO 9001

An Auditor's History of the International Quality Management System Standard

Milton P. Dentch

MJB
Myles-James Books

Published by Myles-James Books, Shrewsbury, Massachusetts

Cover design by Alex Head
Book design by Alex Head

ISBN: 978-1-7348134-0-1: Paperback
ISBN: 978-1-7348134-1-8 E-book - EPUB

Printed in the United States of America.

Contents

Preface

The number of companies in the United States holding ISO 9001 certificates in 2006 was 44,843. By 2018, the number of US companies maintaining ISO 9001 certificates had dropped to 21,848: a decline of over 50 percent.[1] Data from the US Census Bureau indicates that there are approximately 250,000 manufacturing firms in the United States as of 2017. Less than 10 percent (8.7 percent) of these firms currently maintain third-party certification to the quality management system (QMS) standard managed by the ISO (International Organization for Standardization). The acceptance of ISO 9001 on a worldwide basis is somewhat greater than in the United States, although the worldwide total is also trending downward. In 2018, a total of 878,664 worldwide companies held ISO 9001 certificates, down from 1,058,504 in 2017: a reduction of 17 percent.[2] The acceptance of ISO 9001 as a support tool for QMSs is in serious decline.

I became involved with ISO 9001 in 1992 while working for the instant-film and camera company Polaroid. The company needed to become certified to ISO 9001 to satisfy the require-

1 ISO Survey of Management System Standard Certifications, 2018.
2 ISO Survey, 2018.

ments of hospitals in Europe. The origin of ISO's third-party quality-certification process in 1987 roughly aligned with the formation and growth of the European Union. ISO 9001 certification was one of the standardization concepts that EU members thought would help create a level playing field for member nations. After a year or so, Polaroid was certified to ISO 9001, and sales of film and X-ray products continued unrestricted in Europe. I did not observe any quality improvements in Polaroid film related to ISO 9001, as the ISO certification was managed as a "project" with little connection to the day-to-day activities at Polaroid. As director of worldwide materials for Polaroid at that time, I also witnessed many of Polaroid's suppliers working diligently to become certified to ISO 9001, with questionable results.

I left Polaroid in 1996 to establish a consulting business focused on materials management. My first client engaged me to help build a quality system that would comply with ISO 9001 requirements. While I was still somewhat skeptical of the benefits of ISO certification, I realized that the types of companies I hoped to work with would also be looking for assistance in ISO 9001 certification. To enhance my understanding of ISO, I enrolled in a lead auditor training course, and in a short time, I became proficient in ISO auditing. In taking my first client to successful ISO certification, I started to become aware of the benefits of the ISO process. I liked the discipline and consistency in the preparation of specification and work instructions that ISO required. Over the next several years, I focused on auditing: first as an ISO 9000 auditor for quality, and later for ISO 14001 for environmental management systems. I provided third-party audits for two international certification bodies for fifteen years, conducting five hundred audits in total. I progressed from ISO skeptic to ISO supporter. I particularly enjoyed auditing small companies: machine shops, plastic-forming companies, electronics fabricators, and chemical-processing plants. While I often noted issues with the lack of clarity in the ISO 9001 requirements and inconsistent interpretation of the requirements by my auditor colleagues, overall I felt the ISO process was a useful quality tool for manufacturing companies in providing a disciplined approach to quality management. By 2014, I'd started to wind down my auditing time and had begun to focus on auditor training and consulting. For the next several years, I assisted organizations in upgrading to ISO 9001:2015 and ISO 14001:2015. The release of the new ISO 9001:2015 and ISO 14001:2015 standards presented an opportunity for consultants to help organizations upgrade to the new standards.

In 2010, the ISO/TC176/SC2—the ISO's Technical Committee (TC) 176, Subcommittee 2 for Quality Systems—launched a worldwide survey of the users and potential users of ISO 9001. The objective of the survey was "to better understand current and future customer needs for ISO 9001 and ISO 9004 and to ensure their relevance into the future." The results were mostly supportive of the ISO 9001:2008 revision. The major takeaway from the survey, according to TC176/SC2, was as follows.

> The survey responses indicate that ISO 9001:2008 is a good document that is relevant for the future, with some enhancement. Comments made by respondents also suggest major changes are not required, but there are improvements that could be made, and special attention should be given to ensure the correct application of the standard.[1]

1 www.iso.org/tc176/sc2/ISO9000UserSurvey.

The enhancement referenced described concerns for the weak end-user familiarity with ISO 9004:2000 and ISO 9004:2009, the guidance standards for organizations to support the achievement of sustained success through a quality management approach. The TC176/SC2 survey questioned how end users of ISO 9001 were able to integrate the quality standard with ISO 14001, the environmental standard. A third of the responders had a favorable response related to their ability to integrate ISO 9001:2008 with the environmental standard ISO 14001. Chapter 6 in *The Rise and Impending Demise of ISO 9001* provides an overview of the TC176/SC2 user survey and links to the complete report.

The Technical Management Board (TMB) of ISO leadership created Annex SL in 2012. The goal of Annex SL was to harmonize all ISO management systems, terminology, and formatting to facilitate the integration of multiple management systems. If the TMB was aware of the TC176/SC2 user survey, they chose to ignore the ISO 9001 end-user feedback that indicated ISO 9001:2008 was a "good document and relevant for the future." The feedback and justification that drove ISO leadership to create Annex SL has not been publicly shared.

Since the initiation of Annex SL, several management systems standards (MSSs) have been issued under Annex SL requirements, including the quality standard ISO 9001:2015, the environmental standard ISO 14001:2015, and the occupational health and safety (OH&S) standard ISO 45001:2018. Annex SL achieved its overarching goal. The new ISO MSSs now have a common high-level structure, identical core text, and common terms and core definitions. In my opinion, however, the standards created under Annex SL guidance were poorly conceived, formatted, and written by the technical committees. As I became aware of the changes generated by the 2015 versions, I was increasingly disappointed in the major overhauls of ISO 9001:2008 and ISO 14001:2004 generated by the committees charged with creating the new standards. I agreed with the TC176/SC2 end-user survey results that the ISO 9001:2008 standard defined the QMS requirements clearly. The quality management enhancements established by the 2015 version could have been easily tacked on to the ISO 9001:2008 revision without a major overhaul. The changes, including new and questionable requirements and formatting to the environmental standard from ISO 14001:2004, were even more alarming to me. The requirements outlined in the original environmental management version of 1996 and updated in 2004 were also clearly defined. I assisted several organizations in upgrading to ISO 9001:2015 and ISO 14001:2015. During the certification audits, I was dismayed at the confusion and interpretation conflicts the new 2015 standards had placed on my clients.

The initial ISO 9001 standard released in 1987 was built on the principle of "document what you do—do what you document." This concept appeared to me to be too simplistic to be of value to the quality performance of most companies. After providing third-party audits to hundreds of companies, of all types and sizes, I recognized how the discipline required by the ISO 9001 documentation contributed to the consistency necessary for organizations to create products that met their customers' expectations. The ISO 9001 revision in 2000 added *process management* and *customer focus* to the standard, supplementing the consistency requirements. The 2015 ISO 9001 revision diminished or obfuscated the need for clearly defining the specifications and instructions—a serious mistake, in my opinion. Additionally, ISO 9001:2015 injects "risk-based thinking" (RBT) or risk management as a key component of QMS certification—and a replacement for preventive action. ISO 9001 is a quality management support tool. Risk management for an organization is central to the management of the organization's *total business*. The ISO 9001 standard does not provide the structure or auditable requirements to allow assessment of the organization's *business* risks. Quality risk-management requirements can be linked to the *change control* requirements of ISO 9001, as described in chapter 8 of this book.

The Rise and Impending Demise of ISO 9001 is organized into four sections:

- My experiences with the effective application of ISO 9001
- History, organization, feedback processes, and revisions to ISO 9001
- Causes for the decline in ISO 9001 certifications
- Recommendations on the adjustments and management changes required to return ISO 9001 to relevance

In the book, I analyze the 2018 ISO Survey of Management System Standard Certifications to explore trends related to which countries and industry sectors are gaining or losing ISO 9001-certified organizations. I have also sampled a group of ISO 9001-certified companies—the "customers" of the ISO process—to obtain feedback on the pros and cons of ISO 9001:2015. I then use this information to support my recommendations for improving the ISO processes.

Data from the 2018 ISO survey, combined with US census data, indicates that over 90 percent of manufacturing companies in the United States do not find value in third-party certification to ISO 9001. While the worldwide embrace of ISO 9001 certification appears to be greater than in the US, the number of global ISO 9001–certified companies is also trending downward. For the ISO 9001 certification process to stem the loss of certified organizations and regain a position of importance in supporting the QMSs of certified companies, a major adjustment is required in how the MSSs are prepared, managed, and monitored. *The Rise and Impending Demise of ISO 9001* provides my recommendations on the adjustments and management changes required to return ISO 9001 to a position of relevance.

Abbreviations List

AB: accreditation body
ACC: American Certification Corporation
AIAG: Automotive Industry Action Group
APG: Auditing Practices Group [of the ISO]
AS&D: aviation, space, and defense
ANAB: ANSI National Accreditation Board
ANSI: American National Standards Institute
ASQ: American Society for Quality
BOM: Bill of Material
BPR: business process reengineering
BSI: British Standards Institute
BVC: Bureau Veritas Certification
BVQINA: Bureau Veritas Quality International North America
CB: certification body
CQI: Chartered Quality Institute
DIS: draft international standard
EA: economic activity
ECN: engineering change notice
EH&S: environmental health and safety
EMS: environmental management system
FDIS: final draft international standard
FMEA: failure mode and effects analysis
IAF: International Accreditation Forum
IAQC: International Aerospace Quality Group
IATF: International Automotive Task Force
IEC: International Electrotechnical Commission

IRCA: International Register of Certificated Auditors
IS: international standard
ISO: International Organization for Standardization
JTCG: Joint Technical Coordination Group
MAPI: Manufacturers Alliance for Productivity and Innovation
MSS: management systems standard
NACE: Nomenclature des activités économiques dans la Communauté européenne
Nadcap (formerly NADCAP): National Aerospace and Defense Contractors Accreditation Program
NAICS: North American Industry Classification System
NGS: National Graduate School of Quality Management
NMB: National Board Member
NPD: new product development
OEM: original equipment manufacturer
OH&S: occupational health and safety
PPAP: production part approval process
PRI: Performance Review Institute
QMS: quality management system
RAB: Registrar Accreditation Board
RBT: risk-based thinking
RVA: Raad Voor Accreditatie
SDO: standards-developing organization
SIC: Standard Industrial Classification (of economic activities)
TAG: technical advisory group
TC: technical committee
TMB: Technical Management Board
TMC: Technical Management Committee
TQM: total quality management
USNC: United States National Committee

1

ISO 9001 Work and Auditing Experiences

My first experience with ISO occurred in 1992. I was employed as the director of materials for worldwide film manufacturing at the Polaroid Corporation in Waltham, Massachusetts. The company had recently experienced a major challenge: Polaroid instant-film products would no longer be sold in European hospitals unless the film was produced in an ISO 9001–certified plant. Since several million packs of instant film, plus many X-ray camera devices, were sold to hospitals in Europe every year, Polaroid needed to act quickly—and we did. Several engineers and managers were reassigned from their regular jobs to become familiar with the ISO's requirements for quality management in order to assist the film-assembly plants in obtaining certification to ISO 9001. The film-assembly plants achieved certification to ISO 9001 in about a year, and instant-film sales continued at the European hospitals uninterrupted.

Other than maintaining European film sales, what other impact did the ISO certification have on the quality of Polaroid film? From my close observations, I would say very little to none. The folks assigned to implement the ISO certification were not part of the day-to-day operations in the plants; they attacked each of the twenty elements and attendant requirements of ISO 9001 as a short-term project with goals and milestones. Employees attended awareness sessions to be trained on how to answer questions from the ISO auditor. "What is Polaroid's quality policy? Where do you find instructions on how to make film? What steps do you take when film you've made does not meet specifications?" The employees memorized the answers in somewhat rote fashion, and the auditors from Norway were sufficiently impressed during their weeklong series of interviews and investigations to recommend that the plant be certified to ISO 9001. While many nonconformances were raised during the audit, the ISO team convinced the auditors that the film plant had addressed the issues with diligence, and Polaroid customers could be assured that the film produced at these plants would be of *consistent quality*.

In reality, Polaroid instant film, both before and after ISO certification, was plagued with consistent quality problems, inherent in the technology's extremely complex scientific and manufacturing challenges. Customer experience with Polaroid instant-film products through the entire seventy-five years of the products' life cycle indicated that one in four instant pictures were not of high quality. But the perceived value of the instant experience was so great that customers purchased over 100 million packs of film most years until the late 1990s, when the advent of digital imaging essentially made the Polaroid instant-film product obsolete.

During the ISO audit in 1992, managers like me were instructed by the ISO team to stay out of the way—perhaps take a trip or some vacation time that week—as we had not attended the awareness/indoctrination training. As materials director, I was responsible for coordinating the manufacturing planning and scheduling with Polaroid's sales and marketing divisions. In ISO terms, this process was referred to as the "contract review" or the sales-manufacturing forecast. Like many other publicly traded consumer-products companies in the 1990s, Polaroid adjusted the film-sales demand to satisfy Wall Street analysts for the quarter, with little connection to actual consumer sales demand. If the ISO auditor had asked me to explain the contract-review process, Polaroid might never have been certified to ISO 9001! The ISO team selected lower-level employees to answer the auditor's questions using professionally presented spreadsheets that indicated a strong link between sales and manufacturing.

My opinion of the value of ISO 9001 certification, which was shaped by this first experience, continued when I became materials director, as I observed suppliers to Polaroid become certified to ISO without any obvious uptick in quality or delivery performance. The 1990s joke about ISO involved a company that made life jackets; they were manufactured to size specifications, using precise instructions and tight tolerances in an ISO 9001–certified shop. Unfortunately, the technical department selected cement as the material of choice, and the life jackets were useless. The ISO 9001 revision in 2000 provided a solution to the cement life jacket issue by placing more emphasis on results and customer requirements.

Polaroid Retirement

I joined Polaroid in 1969 and had a rewarding twenty-seven-year career at the iconic instant-film company. Starting as a process and plant engineer in the film-coating and chemical plants, I advanced to assignments such as engineering, manufacturing, and technical manager in the Film Coating and Battery Division. In 1989, I was promoted to director of materials for the Integral Film Division. In the materials director's role, I was responsible for maintaining Polaroid's supply chain to provide over 100 million packs of Integral Film annually. Polaroid sourced unique raw materials from suppliers in the US, Europe, and Asia. I gained considerable experience in the selection, purchasing, and quality assurance of a wide range of materials: manganese dioxide from Belgium, PVC from Japan, thin-rolled aluminum foil from North Carolina and Germany. I was also keenly aware of Polaroid's sharply declining sales of instant film due to the rapid rise of digital-imaging options. I accepted an early retirement package in 1996.

When I left Polaroid, I had two goals in mind. I had not completed my master's degree in business when I was at Polaroid (I'd started a few times but had gotten distracted for various reasons), so I wanted to get enough credits to complete my MBA. The second goal was to start a consulting business. I had traveled around the world for Polaroid sourcing materials, so I had a pretty good feel for suppliers, and I had a background in all sorts of paper materials and coatings and batteries and so forth. I thought I had a lot to offer. I also had a few opportunities for jobs in the technical field for former suppliers, but the locations were usually too far away from my home in Massachusetts, where my family was. In 1996, I was offered a consulting project at a local coating company, the Furon Company, in Worcester, Massachusetts. My assignment was to establish a quality system for the company that could be certified to ISO 9001. Several of Furon's customers were pushing Furon to obtain ISO 9001 certification.

National Graduate School of Quality Management

While at Furon, I attended a master's degree program on total quality management. The National Graduate School of Quality Management (NGS), based in Falmouth, Massachusetts, is a weekend program for adults. The curriculum included twelve modules spread out over twelve to fifteen months. The modules covered a wide spectrum of quality subjects, such as statistical process control, Baldrige Award examination, and instruction on the application of a variety of quality tools. One of the last modules in the program was Quality System Certification. I had an option to either take a classroom type of instruction or attend the ISO 9001 lead auditor course. I took the four-day course, which was spread out over a couple of weekends and included a four-hour exam. I passed the Registrar Accreditation Board (RAB) exam, which provided me with the entry-level credentials necessary to conduct third-party audits for ISO 9001. The ISO 9001 lead auditor course instructor was Lorcan Mooney. Lorcan was an Irish citizen and was an early creator of ISO 9001 in Europe.

To qualify to conduct ISO audits, someone who passes the RAB-approved exam has to conduct several audits under the supervision of an approved RAB lead auditor—and then several additional days to qualify as a lead ISO auditor. The challenge is that the potential auditor needs to find a company or registrar that will be willing to provide qualifying audits for a trainee. Fortunately, Lorcan owned a small certification body in Massachusetts called the American Certification Corporation (ACC). After I'd passed the exam in Lorcan's NGS class, he asked some of the class participants if we would like to work with him on a few qualifying audits. Lorcan would be the lead auditor and would sign our audit logs, as he held lead auditor status. I had some vacation time at Furon, so I took a week off and went to Upstate New York, near the Canadian border, with a few classmates and Lorcan. An ACC client had four plants located in the Oswego, New York, area. The plants were part of a multisite wire-drawing company.

Early one Monday morning in 1997, two of my NGS classmates and I drove the 250 miles together from Massachusetts to Oswego. There we would meet Lorcan, who had driven up from Boston the previous day. It was an interesting, enjoyable time. We brought our golf clubs, played nine holes on Monday afternoon, and enjoyed excellent lodging and food at a historic old village inn. We were scheduled to conduct audits in the four plants during the remainder of the week, returning home on Friday. I had some knowledge about the company's process, as I had done drafting work for a steel-producing company while in college; my teammates had previous experience as internal auditors for their previous employers, so we got through the audits fairly well. More importantly, the four audit days provided my teammates and me with a good start to establishing the audit days required to qualify as entry-level auditors.

After I graduated from the NGS, I hired Lorcan to assist Furon in obtaining ISO 9001 certification and to train the internal auditors in the company. Lorcan has an entertaining presentation style and was a big hit with the employees at Furon. He was very helpful in assisting Furon's quest for ISO 9001 certification, which the company achieved after a few fits and starts. During this time, Lorcan and I became good friends. The Irish native had been living and working in Boston for several years, where he managed his ACC company and provided ISO 9001 training. During a golf game one day, Lorcan asked, "If you ever get tired of working at Furon and you want to go back to consulting or something, let me know. You can work with me in my American Certification Corporation."

American Certification Corporation

By 1999, I had spent thirty-five years working an eight-to-five (or longer) workday, five days a week. Although I appreciated the opportunity at Furon, I found being confined in a plant all day kind of restrictive. I wanted to start consulting with a variety of different companies: my post-Polaroid goal. I left Furon and went to work for Lorcan at ACC. I was the only employee of the company. My official title was vice president of certification. My job was to manage the ACC office, provide the client communication and documentation for the ISO 9001 audit reports, and help schedule the ACC auditors. Lorcan had established a network of excellent contract auditors from around the US who would conduct the ISO 9001 audits and forward the audit reports to me. I would then compile the paperwork and forward the audit package to the president of ACC, who would ensure that the audit results and paperwork satisfied the ISO 9001 certification requirements. Gil Andrade, a quality professional and longtime leader of the Boston section of the American Society for Quality (ASQ), was ACC's president. Lorcan and Gil had established a solid certification process that was accredited through the Netherlands' RVA (Dutch for Raad Voor Accreditatie, or accreditation board). RVA accredits conformity-assessment bodies such as ACC. RVA's US equivalent is the American National Standards Institute's (ANSI) National Accreditation Board (ANAB). Lorcan chose RVA based on his experience with them while he worked on ISO 9001 in Ireland and the UK.

At ACC, I scheduled the audits and reviewed the reports. I had to put the reports into a certain format to make sure that they went by the rules, that the audit had been done properly, and that the person who had done the audit had the proper credentials. Two of the auditors were quality engineers who had worked with me at Polaroid. While I was at ACC, an auditor from RVA conducted an office audit of ACC to ensure that we were providing proper ISO 9001 audits. We passed the audit, and in a short period of time, I gained a lot of hands-on experience with the ISO certification and accreditation processes. What I needed personally was to conduct sufficient audits to qualify as, first, a quality auditor, and then as a lead ISO 9001 auditor so that I could work independently for certification bodies.

I scheduled myself as auditor if ACC audits were within driving distance from my home. Since I was not officially an RAB-qualified auditor, I could only conduct surveillance audits, and not full certification or recertification audits. I was approved in ACC's auditor-qualifying procedure by way of Lorcan's witnessing my audits at the New York plant. This is common practice among most certification bodies (CBs), and in fact, many CBs hire auditors and provide them with sufficient auditing experience to qualify them under the CB's procedures: completely bypassing the RAB or equivalent auditor-qualifying bodies, such as the International Register of Certificated Auditors (IRCA). The CB process for self-qualifying auditors is a concern and weakness in the ISO scheme. While many CBs have a disciplined process to ensure that the auditors they employ have the proper credentials and experience to perform audits, opportunities for abuses do exist, particularly when CBs qualify auditors for auditing to the environmental and OH&S standards (ISO 14001 and ISO 45001, respectively). From personal experience, I can attest to the degree of difficulty presented to an experienced environmental professional when applying for approval as an ISO 14001 auditor with RAB (now Exemplar Global).

I added sufficient audit days with ACC in about a year to qualify for quality auditor status with RAB. My next step was to conduct sufficient audits under a lead auditor's supervision to qualify as a RAB lead auditor. I couldn't do that very easily with the number of clients registered by ACC. I came across an advertisement in a local newspaper from a large international certification body called Bureau Veritas Quality International North America (BVQINA). They were looking for auditors in the area, in particular somebody with a paper-mill background. In early 2000, after an interview with BVQINA, in which I described my several years of experience employed as an engineer in the pulp and paper industry, I was hired as a part-time contract auditor. After a few audits under the supervision of BVQINA's lead auditors, I qualified for lead auditor status with RAB and was hired full time by BVQINA. The experience with BVQINA was very important to my new career as an ISO auditor. It was also great to have subsidized health insurance and other benefits again.

Bureau Veritas Quality International North America

The next several years were an amazing and formative time for me. BVQINA's headquarters were located in Jamestown, New York, on the southwestern corner of the state. BVQINA had captured a strong base of process and chemical manufacturing clients all over the United States, with a heavy concentration of clients in Upstate New York and Western Pennsylvania. Most of my previous work life had centered on working directly for process-based manufacturing companies such as Rice Barton (a manufacturer of paper-making machinery, and my first employer), Polaroid, and Furon. (More on process-based and discrete companies later.) I had spent considerable work time as an engineer and manager in paper mills, chemical-manufacturing plants, aluminum-rolling mills, plastic processors, and the like.

I had considerable direct experience working with these industries, so BVQINA was able to qualify me to audit companies with these industry codes. In addition to supporting my qualification for lead quality auditor status, they provided audit experience for me so I could qualify as an RAB environmental ISO 14001 lead auditor. The company was run by Greg Swan. Bureau Veritas was a large, international registrar; Greg had formed the North American division. In my opinion, Greg did an outstanding job organizing the company. He was an auditor himself, so he knew the ins and outs of auditing, and he established policies that treated the staff with consideration. Greg selected experienced industry professionals from companies such as IBM, Corning, and General Motors. BVQINA would minimize the need for its auditors to travel on Sundays or Friday nights.

The company preferred to use auditors who were full-time employees of BVQINA to establish loyalty to the company and provide the audited clients with auditors who had a more consistent approach to interpreting the requirements of the ISO standards. The ISO auditing process has always had the problem of what I refer to as "nomadic" contract auditors, who bounce around from the various CBs, sometimes taking clients with them to their new employer. Many ISO auditors come from a pool of men and women in their sixties and beyond who are looking for a challenging but shortened workweek. To solve this problem, BVQINA allowed auditors to work a three-quarter or half workweek, but they remained company employees with prorated benefits. After three years as a full-time employee, I opted for half-time employment in 2004. It was great to have a consistent part-time audit schedule, with a portion of my medical insurance costs covered.

The auditor-training process at BVQINA was also good. New auditors were witnessed by experienced BVQINA lead auditors before they were allowed to audit independently. Annual auditor exchanges were held in Jamestown (or, in some years, Niagara Falls) that all auditors were obliged to attend to be made aware of issues with auditor interpretations, client complaints, and auditing best practices. The exchanges were also good team-building opportunities. Additionally, BVQINA solicited feedback on the auditor team's performance after the audit report was closed and then shared the feedback with the auditor.

BVQINA also had a very strict process for controlling ISO audits. When assigned to a client, the auditor was required to submit the audit plan at least two weeks prior to the audit to the CB's administrator, who would then forward the plan to the client. This process ensured that the client received the audit plan in a timely fashion and that the CB could forward the audit plan to an alternate auditor, should the assigned auditor become unavailable. This technique also allowed the administrator to monitor the hours spent on the audit to ensure conformance with man-day requirements. BVQINA itself was frequently audited by its accreditation body (AB) and was sensitive to the risks related to approving audits with inadequate audit time. If the AB discovered a deficiency, such as a low number of audit hours at the client's site, then the CB would be obliged to provide a record of the justification for the reduced audit hours, or the CB would have to assign an auditor to return to the client to complete the audit at the CB's expense. If the AB discovered the discrepancy (usually several) during an audit of the CB, then the AB would issue a major nonconformance, which could result in suspension of the CB's accreditation. The shortened audit day's practice was—and is—a major flaw in the overall ISO certification scheme.

BVQINA managed audit time controls by checking the audit plan against their procedures and quote to the client before the audit and held assigned auditors to the plans they submitted. The auditor could still conspire with the client to leave a few hours early; most clients were delighted to say goodbye to the auditor as soon as possible! The BVQINA administrator was aware of this and had an interesting technique to monitor the auditor's time at the client's site: she would check the auditor's travel plan against the audit plan to estimate cycle time from leaving the client to catching a flight. I got caught allegedly leaving an audit too early during a one-day audit in Trenton, New Jersey. I was auditing a distributor. Fifteen minutes before the 4:30 p.m. closing meeting, I called the BVQINA supervisor with a concern I had about the client's certificate scope.

But before I could ask the question, the supervisor, Rob, said, "Milt, you have a 5:30 flight back to Boston. How could you possibly catch that flight in less than an hour?"

"Rob," I responded, "I'm a ten-minute walk from the client's plant to the small Trenton airport, but thanks for your concern."

I had several similar audit assignments at distributors or similar clients who were located either within the airport complex or a few miles away. They were low-stress traveling experiences compared to the many difficult commutes I had. When I returned home that evening, I sent an email suggesting that Rob be more careful when making judgments based on anecdotal information. I received an apology from him.

BVQINA auditors were not allowed to use checklists with prepared audit questions. The thought was that the auditors were sufficiently knowledgeable of the ISO standards that they didn't need the generic list of questions. The audit reports were required to include objective evidence of the audit notes: who was audited, which documents were reviewed, which processes were audited, and what the detailed findings were. BVQINA's technical department would reject audit reports that lacked details. Many CB audit reports I currently observe are quite weak: a copy-and-paste summary loaded with boilerplate.

To ensure impartiality, BVQINA rotated auditors frequently so that a client would not be audited by the same person continually. I have known clients who have been audited by the same auditor for as long as ten years. While there is no rule against doing so for ISO 9001 audits, it really is poor practice; neither the CBs nor the organizations should allow it to occur. The sector-specific management systems (such as automotive and aerospace) are quite restrictive regarding the rotation of auditors, which would be difficult for the CBs to follow for ISO 9001. While auditing for BVQINA, I seldom audited the same client two years consecutively. After leaving BVQINA, when I was assigned a new account as an auditor for other CBs, I would request to only audit a client for no more than four consecutive years. If the client was a first-time-certified client, then I would do the certification audit and the next two surveillances. Ideally, at the recertification audit, I would be joined by an auditor who was new to the client, who would then repeat a similar cycle. While this scenario is ideally how the ISO process should work, the practice is probably not practical due to the difficulty in finding auditors with the proper industry experience qualifications; some clients accepted only auditors who were within driving distance of their site to reduce travel expenses.

Overall, my experience with BVQINA was excellent. I admired their tight controls and professionalism in managing the audits to ensure that their clients received a thorough audit within the rules of ISO accreditation bodies. One requirement is for the CBs to assign third-party auditor teams that have some knowledge of the business process to the organization. An example would be someone who has spent his or her work life providing quality-control activities for an electronics-assembly company. That person, once qualified to be an ISO 9001 third-party auditor, would not be assigned to audit a steel mill or a chemical plant without some additional monitored auditing experience. When a new auditor was qualified, BVQINA would carefully screen that person's résumé to search for work experiences related to various industry sectors, according to the Standard Industrial Classification (SIC) of economic activities. This would provide BVQINA with evidence to support their assigning of the auditor to clients where the auditor had an understanding of the client's manufacturing processes. The CBs are allowed to qualify auditors for industry codes based on experience they've gained of industry-sector knowledge while auditing companies under the supervision of a qualified lead auditor. While this process can be acceptable where the industry sectors are somewhat similar, such as discrete manufacturers (including machining, molding, assembly, and the like, where specific products are made), I don't believe auditors without work experiences in process-based sectors (such as chemicals, paper mills, and basic metals production, where generic products are made) should ever be qualified to independently audit these industries.

I conducted audits for BVQINA for eight years: the first three as a full-time BVQINA employee, and the last five as a part-time contractor. BVQINA used me extensively as an ISO 14001 auditor, often conducting combined ISO 9001 / ISO 14001 audits. I really enjoyed those audits. I conducted audits in a mining quarry, where the company set dynamite charges to remove marble rock to provide the feedstock for manufacturing calcium carbonate at the company's nearby plant that produced calcium carbonate as a filler ingredient for chewing gum and similar materials. I spent three days on an oil rig platform, 150 miles out in the Gulf of Mexico, conducting an environmental audit for a large oil company. I conducted combined ISO 9001 and ISO 14001 audits at a large paper mill in Arkansas as well as at the Crane Paper Mills in Dalton, Massachusetts, the company that has produced the paper for US currency since the Revolutionary War. Really an amazing eight years. I worked with some terrific auditors. I especially enjoyed auditing and traveling with my colleagues, who were also qualified to audit both quality- and environmental-certified clients, often in complex chemical-processing operations. Several of the auditors had similar environmental work experiences as I did, but they had considerably more compliance/regulatory knowledge than me. I learned from them.

But the extensive travel started to wear me down. Some weeks, I would start my audit schedule in Rochester, New York. I'd audit a client there for one or two days and then fly to Newark, New Jersey, and do an additional few days there. The next week, it might be a similar schedule but reversed, starting in New Jersey and ending in Upstate New York. The administrators and schedulers at BVQI were great. They worked with me to try to arrange my schedule to spend four consecutive days (say in Erie, Pennsylvania) auditing two or three small companies.

TÜV SÜD America

In 2006, I noted an ad in my local newspaper by the Massachusetts office of TÜV SÜD, the international certification company, headquartered in Munich, Germany. TÜV SÜD provides auditing, system certification, inspection, and product certification services with locations throughout the world. I was hired by TÜV SÜD America and worked for them for eight years. Their headquarters was in Peabody, Massachusetts, very near my home. They had over 250 clients within driving distance of my home, some as close as five miles. As with BVQINA, TÜV had a lot of great folks in the office who scheduled my work and a collection of terrific auditors who remain friends. I found TÜV to be professionally managed during my time with them. They had controls similar to BVQINA's in qualifying auditors based on industry experience, and their technical-review board was diligent in reviewing audit reports to ensure that auditors were not consulting, or providing "soft" audits (where the auditor does not fully address the audited company's lack of conformance to ISO 9001 requirements).

One great experience I had during my last few years with TÜV was being assigned to the international audit team for a large contract manufacturer, with more than fifty contract sites around the world. (Contract manufacturers assemble and tests products designed by large consumer-based suppliers). This particular contract manufacturer produces components for several television, cell phone, and computer companies. As part of that team, I was one of the US representatives who conducted audits in Ukraine, Poland, Brazil, and Mexico, as well as in US plants. TÜV used me to train their auditors when the ISO standards were revised, and I conducted a few webinars for them that covered both ISO 9001 and ISO 14001 implementation guidance.

While I enjoyed working with both BVQINA and TÜV SÜD America, and I respected their certification processes, I have not audited with either company in the past several years and cannot comment on their current performance. BVQINA has reorganized under new ownership and is now known as Bureau Veritas Certification (BVC).

2

ISO 9001 History and Chronology

The concept of an international standard (IS) for manufacturing started in Europe in the 1980s, led by initiatives from the United Kingdom, namely the British Standard, BS 5750, which provided requirements for companies engaged in quality assurance, production, installation, inspection, and testing. The European Union was developing during that time, and it encouraged more trade among European countries. The need for product standards soon became apparent. ISO 9000 essentially replaced BS 5750 in 1987 and became a worldwide, auditable standard for manufacturing and servicing. Initially, ISO 9000:1987 included three subsets:

> ISO 9001:1987—quality-assurance requirements for companies engaged in design, development, production, installation, and servicing
>
> ISO 9002:1987—quality-assurance requirements for companies engaged in production, installation, and servicing (basically, the same requirements as ISO 9001 but without the design of new products)

ISO 9003:1987—quality assurance in final inspection and tests, covering only the final inspection of finished product without concern for how the product was produced

The intention of ISO 9002 was to allow companies making products (or providing a service) based on their customers' designs to certify to ISO without addressing the design function. Unfortunately, many companies that actually provided design services to their customers decided to finesse the system by not submitting their design process to third-party assessment by the ISO registrar's auditors. This issue was resolved with the revision in 2000, which eliminated ISO 9002, requiring all companies to seek certification to ISO 9001. A company that did not design the products sold under their company name would now have to verify to the third-party auditor that the products they manufactured were designed by their customers. The inspection standard (ISO 9003) was also made obsolete in 2000 and is currently included in ISO 9001. The ISO has a requirement to upgrade its management standard every five years. During the twenty-eight-year period from 1987 to 2015, ISO 9000 progressed through four revisions: ISO 9000:1994, ISO 9000:2000, ISO 9000:2008, and ISO 9000:2015 (the current version), released in September 2015. A review of the previous revisions of ISO 9001 provides the evolutionary—sometimes revolutionary—transformation of ISO 9001. The original twenty elements of ISO 9001 are listed in Table 2.1.

These elements were maintained in the 1994 revision, with emphasis on specifying more clarity in what documentation was required by the certified company. Unfortunately, organizations tended to interpret these requirements by creating even more procedures and documentation, often with little value added to the company's quality performance. The essence of both ISO

9001:1987 and ISO 9001:1994 was "document what you do—do what you document," where procedures were seen to be more important than results or improved quality. The ISO 9001:2000 revision was a somewhat radical move for ISO 9001, in that now the focus was on customer satisfaction, process management, and continual improvement. The ISO 9001:2008 revision had no new requirements, just a continued emphasis on process management and customer satisfaction.

4.1	Management responsibility
4.2	Quality system
4.3	Contract review
4.4	Design control
4.5	Document and data control
4.6	Purchasing
4.7	Control of customer-supplied product
4.8	Product identification and traceability
4.9	Process control
4.10	Inspection and testing
4.11	Control of inspection, measuring and test equipment
4.12	Inspection and test status
4.13	Control of nonconforming product
4.14	Corrective and preventive action
4.15	Handling, storage, preservation & delivery
4.16	Control of records
4.17	Internal quality audits
4.18	Training
4.19	Servicing
4.20	Statistical techniques

Table 2.1: ISO 9001:1987 Elements

The current revision (ISO 9001:2015) adds requirements to make (or attempt to make) ISO 9001 a major driver in the business model of the organization as well as added requirements for the organization to demonstrate integration of the QMS requirements into the organization's business processes. The 2015 revision also requires the organization to provide risk analysis in support of meeting the quality objectives and fulfilling process improvements. Additionally, ISO 9001:2015 requires organizations to consider the external issues and interested parties relevant to the QMS other than traditional customers, suppliers, and employees. External issues, such as new technology, potential market forces, and competition that could affect the organization's business strategy, are open to auditing in the ISO 9001:2015 scheme.

3

ISO ORGANIZATIONAL STRUCTURE

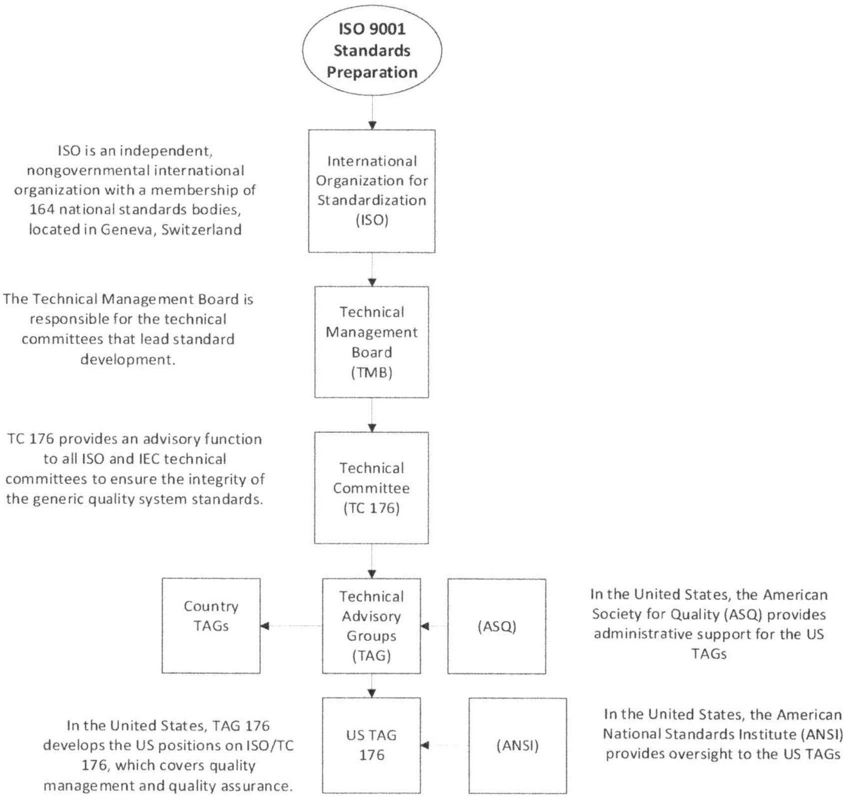

Figure 3.1: ISO 9001 Standards Preparation

Several different organizations and groups create and manage MSSs. The chart in Figure 3.1 provides an overview of these groups. Figure 3.2 outlines the organizations and groups that provide oversight to the management system CBs and third-party auditors.

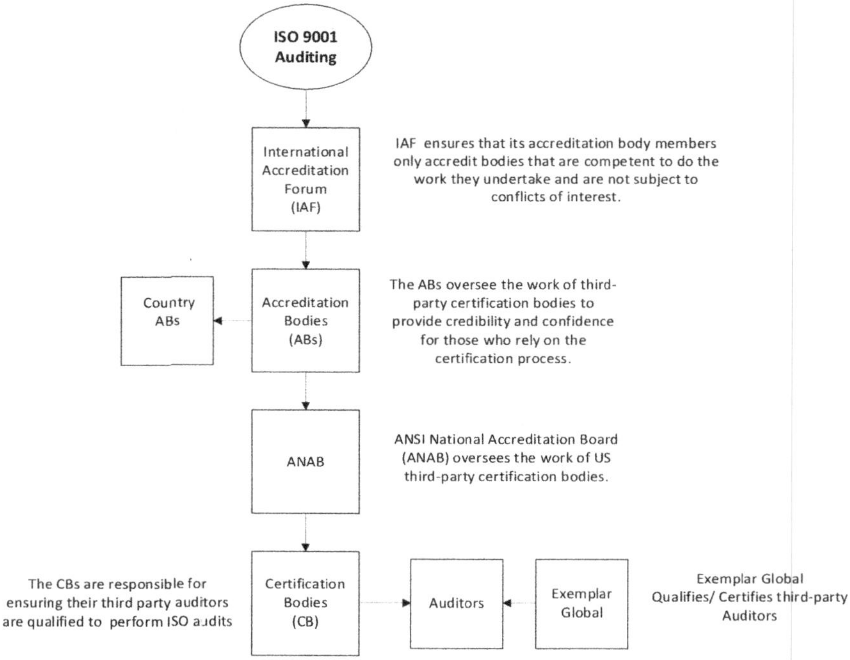

Figure 3.2: Oversight of Third-Party Auditing

The ISO is an independent, nongovernmental international organization with a membership of 164 national standards bodies. Through its members, the ISO brings together experts to share knowledge and develop voluntary, consensus-based, market-relevant ISs that support innovation and provide solutions to global challenges. The ISO story began in 1946, when delegates from twenty-five countries met at the Institute of Civil Engineers in London and decided to create a new international organization to facilitate the international coordination and unification of industrial standards. On February 23, 1947, the new organization, the ISO, officially began operations. Since then, the ISO has published over twenty-two thousand ISs, covering almost all aspects of technology and manufacturing. Today, the ISO has members from 164 countries and 779 technical committees and subcommittees to take care of standards development. More than 135 people work full time for ISO's central secretariat in Geneva, Switzerland.[1]

The Technical Management Board (TMB). The technical work is carried out under the overall management of the TMB. The TMB reports to the ISO Council, and its role is defined in the statutes of the organization. Specifically, it is responsible for tasks such as setting up the various technical committees (TCs), appointing TC chairs, and monitoring the progress of the technical work. The TMB is also responsible for the directives, which are essentially the rules for the development of international standards (ISs), and the board deals with all matters of strategic planning, coordination, performance, and monitoring of TC activities.[2]

Technical Committees (TC 176). The technical committees are responsible for standardization in the field of quality management (generic QMSs and supporting technologies) as well as quality-management standardization in specific sectors at the request of the affected sector and the ISO TMB. ISO/TC 176 is also entrusted with an advisory function to all ISO and International Electrotechnical Commission (IEC) technical committees to ensure the integrity of the generic quality system standards and the effective implementation of the ISO/IEC sector policy on QMS deliverables. In the United States, US TAG 176 (technical

1 To learn more about the ISO structure and activities, visit https://www.iso.org.
2 ISO.org.

advisory group) develops the US positions on ISO/TC 176, which covers quality management and quality assurance.

The International Accreditation Forum (IAF). The IAF is the world association of Conformity Assessment Accreditation Bodies and other bodies interested in conformity assessment in the fields of management systems, products, services, personnel, and other similar programs of conformity assessment. Its primary function is to develop a single worldwide program of conformity assessment which reduces risk for business and its customers by assuring them that accredited certificates may be relied upon. Accreditation assures users of the competence and impartiality of the body accredited. Accreditation is the independent evaluation of conformity assessment bodies against recognized standards to ensure their impartiality and competence. Through the application of national and international standards, government, procurers, and consumers can have confidence in the calibration and test results, inspection reports, and certifications provided. Accreditation bodies are established in many countries with the primary purpose of ensuring that conformity assessment bodies are subject to oversight by an authoritative body.[1]

Accreditation Bodies (ABs). The ABs oversee the work of third-party certification bodies to provide credibility to the certification process for MSSs such as ISO 9001 and ISO 14001. In the United States, the ANSI National Accreditation Board (ANAB) oversees the work of US third-party certification bodies.

Certification Bodies (CBs). The certification body (also called an accredited registrar) is an organization accredited by a recognized body for its competence in auditing and issuing certification to confirm that an organization meets the requirements of a standard (e.g., ISO 9001 or ISO 14001). Accreditation means that certification and inspection bodies have been assessed against

1 IAF.nu

recognized standards to demonstrate their competence, impartiality, and capability.

American National Standards Institute (ANSI). ANSI facilitates the development of American National Standards (ANSs) by accrediting the procedures of standards-developing organizations (SDOs). These groups work cooperatively to develop voluntary national consensus standards. Accreditation by ANSI signifies that the procedures used by the standards body in connection with the development of ANSs meet the institute's essential requirements for openness, balance, consensus, and due process.[1] In the US, ANSI provides oversight to the US TAGs. Through ANSI, the US has immediate access to the ISO and IEC standards-development processes. ANSI participates in almost the entire technical program of both the ISO and the IEC, and it administers many key committees and subgroups. Part of its responsibilities as the US member body to the ISO include accrediting US TAGs, whose primary purpose is to develop and transmit, via ANSI, US positions on activities and ballots of the international technical committee. US positions for the IEC are endorsed and closely monitored by the United States National Committee (USNC) Technical Management Committee (TMC).

American Society for Quality (ASQ). The American Society for Quality is the world's leading authority on quality. The principles of quality influence manufacturing and other industrial processes by focusing attention on how products and services define and meet technical specifications. Quality standards from the ASQ advance learning, quality improvement, and knowledge exchange to improve business results, and to create better workplaces and communities worldwide. Headquartered in Milwau-

1 ansi.org.

kee, Wisconsin, ASQ operates regional centers in North Asia, South Asia, Latin America, the Middle East, and Africa. [1]

Exemplar Global. Exemplar Global provides personnel certification, credential management for people seeking further professional recognition, and independent certification for training providers to enhance their course quality and outcomes. Exemplar Global commenced quality certifications in the United States and Australia in 1989. The company was created in 2004 from the merger of the personnel-certification activities of the US Registrar Accreditation Board (RAB) with the Australia-based Quality Society of Australasia.[2]

The International Register of Certificated Auditors (IRCA). The IRCA, a division of the CQI (Chartered Quality Institute), also provides certification for ISO management system auditors throughout the world.[3]

1 asq.org
2 exemplarglobal.org.
3 quality.org

4

ISO 9001 Revisions and Annex SL

As noted earlier, the ISO has a requirement to upgrade the management standards every five years. From 1987 to 2015, ISO 9001 progressed through four revisions. During the revision of MSSs, the TAGs in the member countries, along with the ISO central technical committees (TCs), generate drafts of the revised standard for review, normally in two drafts: the draft international standard (DIS) and the final draft international standard (FDIS). In the years prior to the 2015 revision of ISO 9001, the ISO created Annex SL. The ISO's TMB replaced Annex SL in May 2019 with Annex L. "Annex L" refers to the annex in the ISO/IEC directives, part 1, 2019 edition, which deals with proposals and criteria for MSSs. It was introduced in 2012 as Annex SL of the ISO supplement to the ISO/IEC directives and was originally only applicable to MSSs developed by the ISO. Starting in 2019, the IEC also applies this annex.[1] The revised Annex L does not appear to provide any new guidance on how TC 176 will prepare future ISO 9001 MSSs.[2]

---

1 ISO/TC176.

2 Annex SL may be accessed at http://users.neo.registeredsite.com/9/2/0/18247029/assets/Annex_SL.pdf.

In 2012, the Technical Management Board (TMB) of the ISO commissioned the Joint Technical Coordination Group (JTCG) to develop Annex SL. All MSSs issued after 2012 are required to be structured in conformance to ten high-level clauses:

Clause 1: Scope

Clause 2: Normative references

Clause 3: Terms and definitions

Clause 4: Context of the organization

Clause 5: Leadership

Clause 6: Planning

Clause 7: Support

Clause 8: Operation

Clause 9: Performance evaluation

Clause 10: Improvement

The ISO technical committees that develop MSSs are required to include the same high-level structure and identical core text, common terms, and core definitions. The belief shared by ISO management and ISO CBs is that with Annex SL in place, organizations will experience less duplication and confusion when implementing multiple MSSs—and the third-party management system auditors will be more efficient when using a common set of requirements across the various standards and industry sectors.

In the several years since the initiation of Annex SL, several new MSSs have been issued under Annex SL requirements, including the quality standard ISO 9001:2015, the environmental standard ISO 14001:2015, and the OH&S standard ISO 45001:2018. Annex SL achieved its overarching goal. The new ISO MSSs do now have a common high-level structure and other factors noted above. In my opinion, however, the standards created under Annex SL guidance were poorly conceived, formatted, and written by the technical committees chartered to create the standards. This book details how the initiation and implementation of Annex SL guidance caused serious defects in ISO 9001:2015, thus reducing the clarity of the requirements of the standard.

5

ISO 9001:2008 End-User Survey: TC176

According to the introduction to the ISO's 9001:2008 end-user survey,

> During its February 2009 meeting in Tokyo, Japan, ISO/TC176/SC2 ... identified the need to conduct a world-wide survey of the users and potential users of its standards. The objective of the survey was to better understand current and future customer needs for ISO 9001 and ISO 9004 and to ensure their relevance into the future. A team was established and preparatory work began in July 2009. An online survey was launched in October 2010 and responses were accepted until the end of February 2011.[1]

To my knowledge, this is the most recent (and only) survey ever conducted by a group under ISO management to acquire feedback from the end users of ISO 9001: the customers. It was

1 https://www.qualitydigest.com/inside/standards-news/iso-survey-seeks-feedback-iso-9001-users-111510.html. https://standards-stores.com/downloads/iso-9001-2008-User-Survey.pdf.

a comprehensive, professionally organized survey that utilized the popular web-based Survey Monkey tool for data collection. A total of 11,722 survey responses were received from 122 countries, submitted in eleven languages. While the number of responses was high, it represented less than 1.5 percent of the number of worldwide ISO 9001 certificates—and the responders were a mixture of ISO 9001-certified companies and individuals (consultants and registrars). I believe the inclusion of responders other than end users in the survey may have distorted the results, as consultants and registrars tend to view ISO 9001 in a mostly favorable bent; ISO 9001 certifications generate revenue for consultants and registrars.

The results were mostly supportive of the ISO 9001:2008 revision. The major takeaway from the survey according to TC176/SC2 was as follows:

> The survey responses indicate that ISO 9001 is a good document that is relevant for the future, with some enhancement (see 5.4 and 5.5 above). Comments made by respondents also suggest major changes are not required, but there are improvements that could be made, and special attention should be given to ensure the correct application of the standard.

The enhancement referenced described a lack of familiarity with ISO 9004:2000 and ISO 9004:2009, the guidance standards for organizations to support the achievement of sustained success by a quality-management approach. Two noteworthy charts in the survey's report described the responders' perception of the benefits and relevance of ISO 9001 (see Figures 5.1 and 5.2).

It is interesting to note that the TMB of ISO Geneva initiated work on Annex SL during the same time frame, based on the TMB's assumption (or feedback) that all MSSs needed to be redesigned to facilitate the integration of multiple MSSs. The TC176/SC2 survey questioned how end users of ISO 9001 were able to integrate the quality standard with the environmental standard, ISO 14001. A third of the responders favorably indicated the integration of ISO 9001:2008 and the environmental standard ISO 14001.

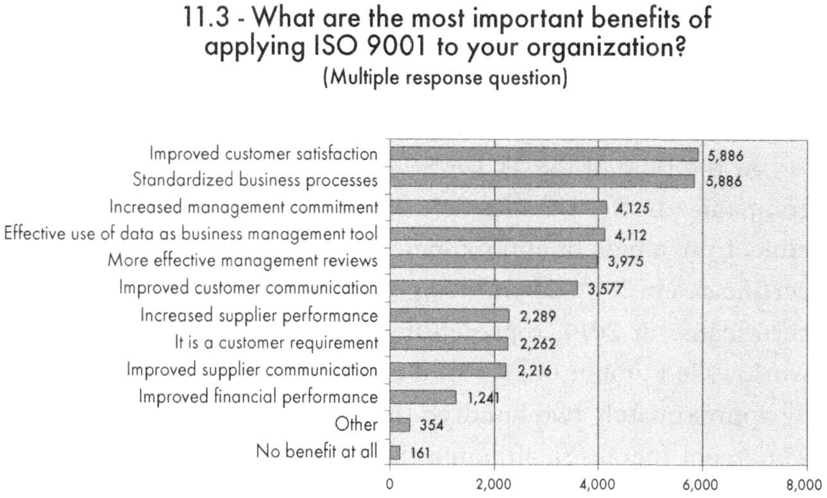

Figure 5.1: ISO 9001 Organizational Benefits

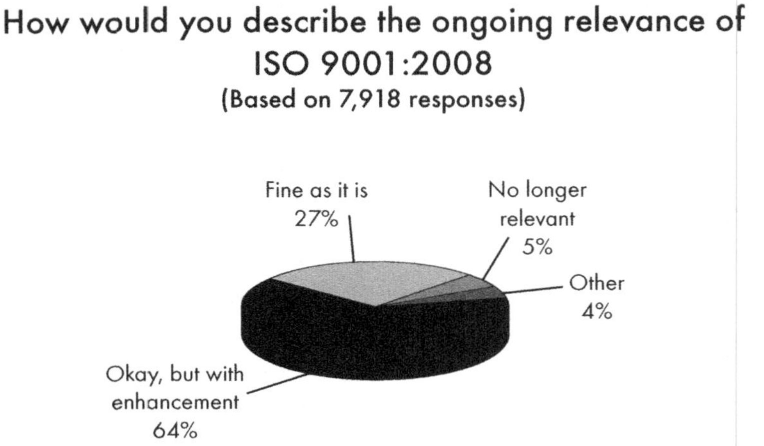

Figure 5.2: ISO 9001 Relevance

At the time of the TC176/SC2 survey in 2010, the number of companies in the US that held ISO 9001 certificates was in decline, from a high of approximately forty-five thousand ISO 9001 certificates in 2006 down to approximately twenty-five thousand certificates in 2010, representing a 22 percent reduction. The worldwide number of ISO 9001 certificates had risen since 2006 by approximately two hundred thousand ISO 9001 certificates (a 25 percent increase), although the ISO survey numbers are now in dispute due to confusion over the number of certified sites versus certified companies.

Was TMB aware of the TC176/SC2 user survey and chose to ignore the ISO 9001 end-user feedback? The feedback to the TMB that drove Annex SL has not been publicly shared in any ISO.org public communication that I may access. Annex SL created the guidelines that resulted in the flawed ISO 9001:2015 standard, in conflict with what the ISO 9001 end-user TC176/SC2 survey indicated.

6

The ISO Survey of Management System Standard Certifications (2018)

The ISO Survey of Management System Standard Certifications is an annual survey of the number of valid certificates to ISO MSSs worldwide. The providers of data are the accredited CBs. The data point for 2018 is not consistent with the previous years' data. Reporting prior to 2018 included a mixture of the number of *certificates* and the number of *sites* that were operating under a company's certificate. A certificate is the document issued by a certification body once the client has demonstrated conformity to the standard, while a site is a permanent location where an organization carries out work or a service. Many large organizations have several of their sites (or plant locations) covered by one certificate. The ISO 2018 survey describes this abnormality as follows, under a section titled "Suppression of Data on 'Sites'":

> Past survey results included data on number of "sites" in addition to the number of "certificates." The goal was to show the coverage of certification by including data on the number of sites that were covered by all the certificates reported. However, the data collected was incomplete and thus often not reported, leading to inconsistent results for many countries. Based on this, it was decided to stop the collection of data on the number of "sites" and to only show the number of valid "certificates."

For 2018, the ISO reported 21,848 certificates held by US companies, with 31,546 sites in the United States. In 2006, the ISO listed 44,883 US companies holding ISO 9001 certificates: a reduction of 51 percent. Figure 6.1 tracks the history of US ISO 9001–certified countries.

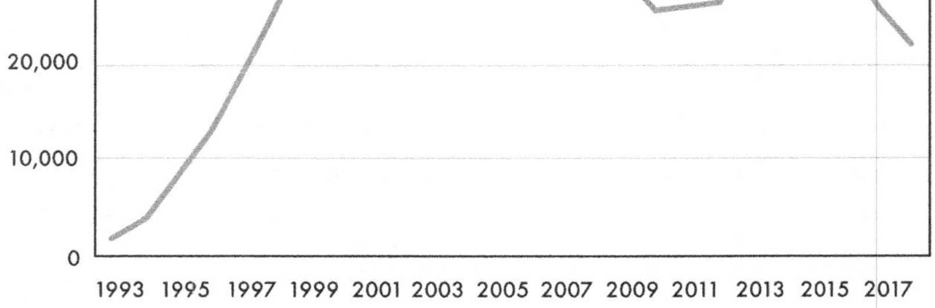

Figure 6.1: ISO 9001 Certificates, US, 1993–2017

Number of Manufacturing Companies in the United States

There are currently approximately 250,000 companies in the United States in the manufacturing sector.[1] Less than 10 percent (21,8848) of US manufacturing companies are currently certified to ISO 9001:2015, which indicates the minor impact ISO 9001 certification currently has on the quality of American-made products.

ISO 9001-Certified Companies Worldwide

The acceptance of ISO 9001 on a worldwide basis is somewhat greater than in the United States, although the worldwide number of ISO-certified companies is also trending downward. The ISO reports that 878,664 worldwide certificates were maintained in 2018, while 1,058,504 ISO 9001 certificates were reported in 2017. The worldwide total dropped by 179,840 certificates for a 17 percent reduction (see Figure 6.2).

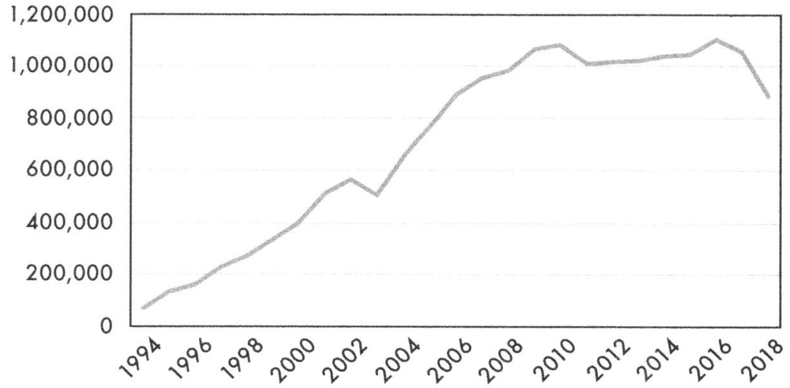

Figure 6.2: ISO 9001 Certificates: Worldwide, 1994–2018

1 https://www.nam.org/facts-about-manufacturing/.

Table 6.1 lists the top thirty-three countries based on sites certified to ISO 9001 in 2018. Figure 6.3 illustrates the number of ISO 9001 certificates/sites by country, sorted to include the countries with more certified companies than the United States. What is interesting is the large number of countries that hold many more ISO 9001 certificates than the United States does. Seven countries currently hold more ISO 9001 certificates than the United States: China, Italy, Germany, Japan, Spain, the United Kingdom, and India. China leads the way, with over 295,000 firms certified to ISO 9001. With the exception of China, the data appears to be confounded by the mixture of certificates and sites in the tabulation of the data.

COUNTRY	CERTS	COUNTRY	CERTS
China	295,703	Thailand	8,401
Italy	87,794	Turkey	7,988
Germany	47,482	Israel	7,845
Japan	34,335	Netherlands	7,841
India	31,795	Switzerland	7,833
Spain	29,562	Australia	6,672
United Kingdom	26,434	Hungary	6,658
United States	**21,848**	Portugal	6,623
France	21,095	Mexico	6,535
Brazil	16,351	Argentina	6,198
Korea	14,123	Greece	6,165
Czech Republic	11,740	Bulgaria	5,943
Poland	11,294	Indonesia	5,714
Colombia	10,027	Canada	4,894
Malaysia	9,558	Belarus	4,663
Romania	9,299	Russia	4,497
Taiwan	8,553	Slovakia	4,012

Table 6.1: Top ISO 9001 Certificates by Country

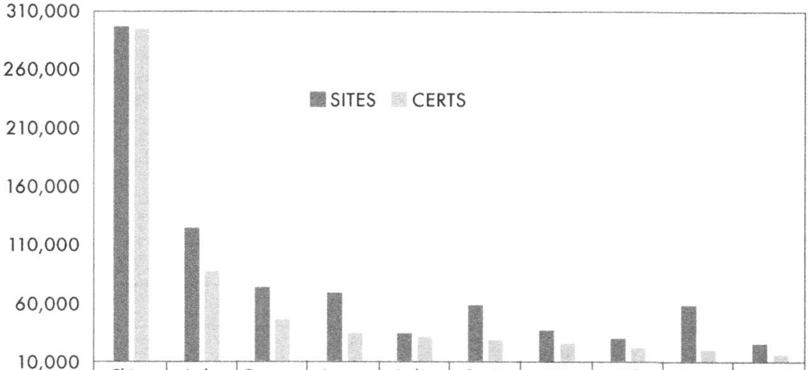

	China	Italy	German	Japan	India	Spain	UK	US	France	Brazil
SITES	297,437	125,994	73,559	68,508	33,947	59,425	36,950	31,546	58,467	25,702
CERTS	295,703	87,794	47,482	34,335	31,795	29,562	26,434	21,848	21,095	16,351

Figure 6.3: ISO 9001 Certificates by Country

ISO 9001-Certified Companies by Industry Sector: Worldwide

The ISO survey includes a breakdown of the ISO 9001:2015-certified companies by industry codes. Tables 6.2A and 6.2B list the number of certificates issued by industry sector. The ISO uses the European industry standard classification system known as NACE (for Nomenclature des activités économiques dans la Communauté européenne). It is similar in function to Standard Industry Classification (SIC) and the North American Industry Classification System (NAICS) for classifying business activities. The twenty NACE codes with the most worldwide numbers of ISO 9001:2015 certificates are ranked in Figure 6.4.

Code	Sector	Certs
1	Agriculture, fishing, and forestry	3,189
2	Mining and quarrying	3,455
3	Food products, beverages, tobacco	23,182
4	Textiles and textile products	10,930
5	Leather and leather products	1,967
6	Manufacture of wood	4,381
7	Pulp, paper, and paper products	8,372
8	Publishing companies	380
9	Printing companies	7,416
10	Manufacture of coke, petroleum	1,595
11	Nuclear fuel	170
12	Chemicals, chemical fibers	28,426
13	Pharmaceuticals	2,960
14	Rubber and plastic products	42,044
15	Nonmetallic mineral products	9,894
16	Concrete, cement	10,897
17	Basic metal and fabricated metal	100,430
18	Machinery and equipment	59,422
19	Electrical and optical equipment	76,861
20	Shipbuilding	1,859

Table 6.2A: ISO 9001 Certificates by Industry Sector (Code 1-20)

Code	Sector	Certs
21	Aerospace	1,397
22	Other transport equipment	9,913
23	Mfg. not elsewhere classified	11,238
24	Recycling	2,857
25	Electricity supply	2,134
26	Gas supply	659
27	Water supply	1,521
28	Construction	75,080
29	Wholesale and retail	78,769
30	Hotels and restaurants	4,429
31	Transport, storage	24,505
32	Financial	9,234
33	Information technology	35,734
34	Engineering services	41,922
35	Other services	47,023
36	Public administration	4,434
37	Education	13,459
38	Health and social work	15,556
39	Other social services	9,208
	Sector not known	85,796
	Total Certificates	**878,664**

Table 6.2B: ISO 9001 Certificates by Industry Sector (Code 21-39)

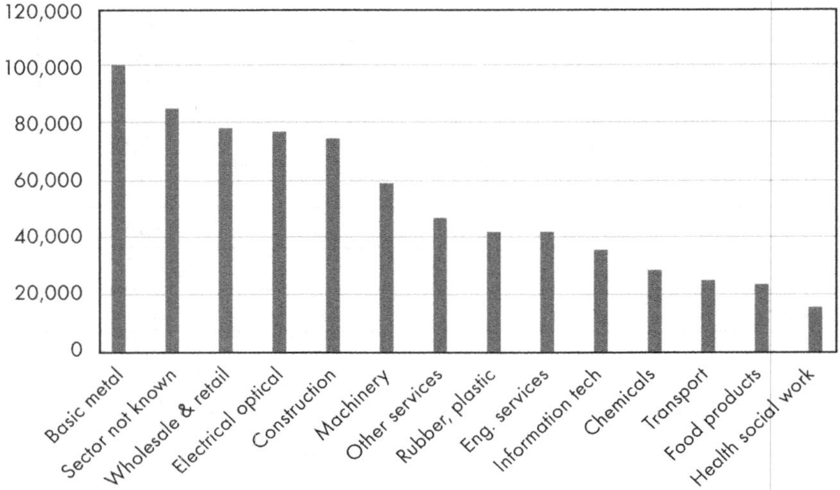

Figure 6.4: Top Fifteen Sector Codes with ISO 9001:2015 Certificates

To further understand the industry classifications that embrace ISO 9001 certification, I divided the thirty-nine NACE codes by companies that manufacture products and those that provide services (see Tables 6.3A and 6.3B and Figure 6.5). The 2018 ISO survey data compiled based on industry sector indicates that about half the worldwide companies that currently maintain ISO 9001:2015 certificates are engaged in manufacturing products or provide materials for products. The remaining ISO 9001-certified companies are providers of some level of services, many of which support the manufacturing companies or the sale and distribution of manufactured products.

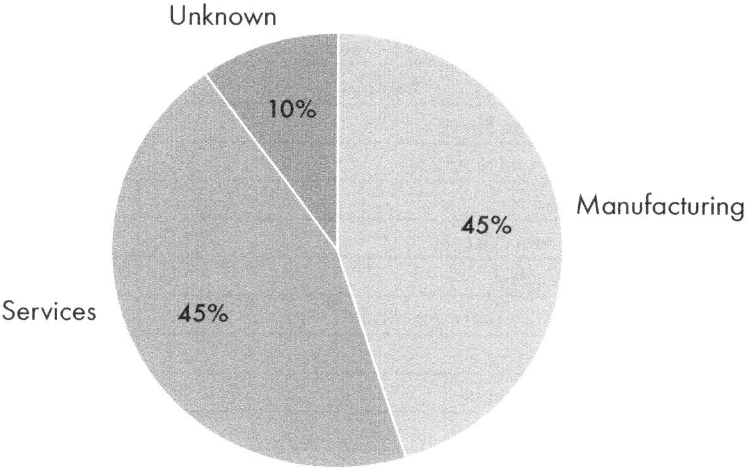

Figure 6.5: Worldwide ISO 9001:2015 Sectors: Manufacturing and Services

Manufacturing		
Code	**Sector**	**Certs**
3	Food, beverages, tobacco	23,182
4	Textiles and textile products	10,930
5	Leather and leather products	1,967
6	Manufacture of wood	4,381
7	Pulp, paper, paper products	8,372
8	Publishing companies	380
9	Printing companies	7,416
12	Chemicals, chemical fibers	28,426
13	Pharmaceuticals	2,960
14	Rubber and plastic products	42,044
15	Nonmetallic mineral products	9,894
17	Basic metal and fabricated metal	100,434
18	Machinery and equipment	59,422
19	Electrical and optical equipment	76,861
21	Aerospace	1,397
23	Mfg. not elsewhere classified	11,238
	Manufacturing Total	**389,304**
	Sector not known	**85,796**

Table 6.3A: NACE Sector with Manufacturing Codes

Services		
Code	**Sector**	**Certs**
1	Agriculture, fishing	3,189
2	Mining and quarrying	3,455
11	Nuclear fuel	170
16	Concrete, cement	10,897
20	Shipbuilding	1,859
22	Other transport equipment	9,913
24	Recycling	2,857
25	Electricity supply	2,134
26	Gas supply	659
27	Water supply	1,521
28	Construction	75,080
29	Wholesale and retail	78,769
30	Hotels and restaurants	4,429
31	Transport, storage	24,505
32	Financial	9,234
33	Information technology	35,734
34	Engineering services	41,922
35	Other services	47,023
36	Public administration	4,434
37	Education	13,459
38	Health and social work	15,556
39	Other social services	9,208
	Service Total	**396,007**

Table 6.3B: NACE Sector with Servicing Codes

ISO 9001-Certified Companies for Construction and Wholesale: China and Europe

In the last several years, the CBs in China and several European countries have granted thousands of ISO 9001 certificates to companies engaged in construction and wholesale/retail activities. In my review of the ISO Survey of Management System Standard Certifications data, I was surprised to see the number of service-based companies, such as construction and wholesale/retail companies, outside the US certified to ISO 9001. I have witnessed only a handful of US companies engaged in construction or wholesale/retail that found application of ISO 9001 certification to be of value to their business. Chapter 8 describes my experiences with service-based companies related to the application of ISO 9001. Figure 6.6 ranks the number of construction and wholesale companies for several countries.

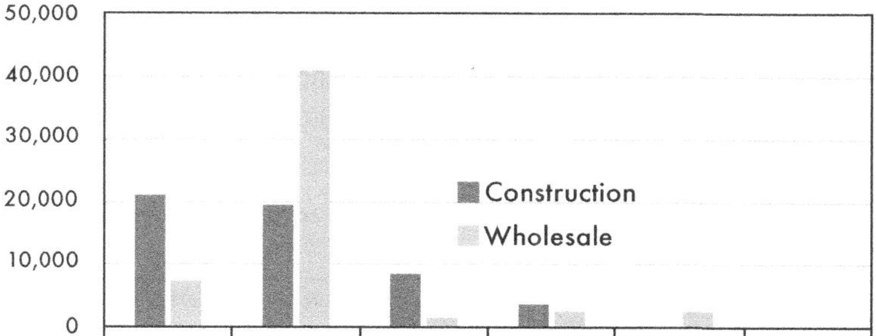

	Italy	China	Japan	Spain	Germany	United States
Construction	20,816	19,210	8,403	4,034	494	22
Wholesale	7,311	40,964	1,552	2,802	2,703	691

Figure 6.6: ISO 9001 Certificates: Construction and Wholesale

US ISO 9001 Certificates by Service Sector

The embrace of ISO 9001 certification by service-related companies has not occurred in the United States. Of the 21,848 US companies holding ISO 9001 certificates, 2,248 (10 percent) of the certificates were issued to service-based companies (Figure 6.7). For reference, Figure 6.8 illustrates the ISO 9001 certificates for the US industrial sectors with greater than two hundred certificates.

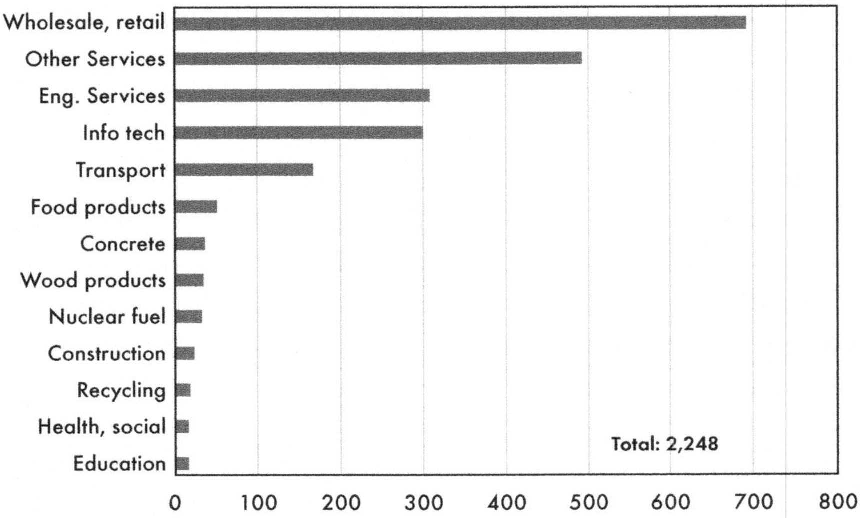

Figure 6.7: ISO 9001 Certificates: United States, by Service Sector, 2018

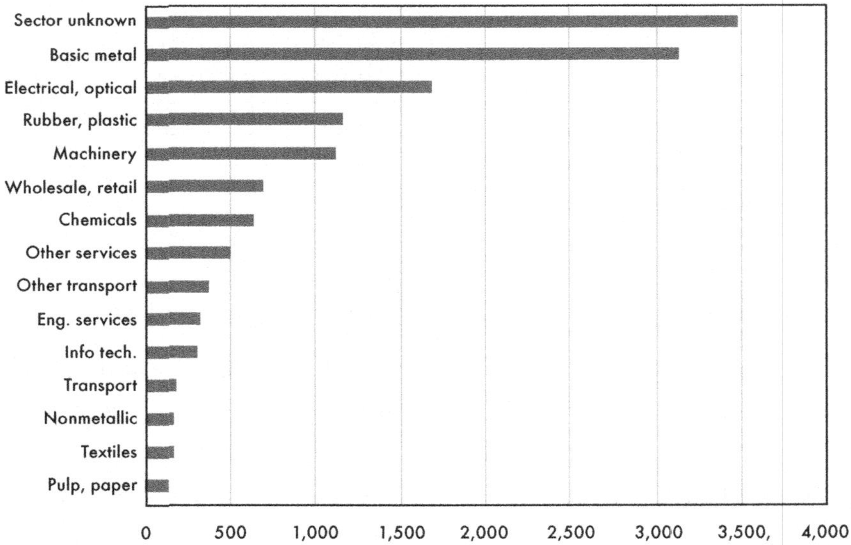

Figure 6.8: ISO 9001 Certificates: United States, Largest Sectors, 2018

ISO 9001 Certificates Issued without Industry-Sector Classification

A key component of a CB's guidance for the issuing of a certificate is its verification of the scope of the company's processes, with industry-code classification confirmed to match the scope. The industry sector provides the CB with evidence that the audit days assigned are consistent with the auditee's manufacturing or service activities (and the assigned auditor's qualifications). The 2018 ISO survey data reports that considerable ISO 9001 certificates have been issued in several countries without identification of industry-sector code (Table 6.4 and Figure 6.9).

Country	Unknown Sector	ISO 9001 Certificates	% Unknown Sector
Israel	5,699	7,876	72%
Germany	18,086	47,482	38%
UK	8,755	26,434	33%
US	3,487	21,848	16%
Japan	4,099	34,335	12%
Italy	2,030	87,794	2%
China	4,286	295,703	1%
Total	**46,442**	**521,472**	**9%**

Table 6.4: ISO 9001 Certificates by Country without Industry-Sector Designation

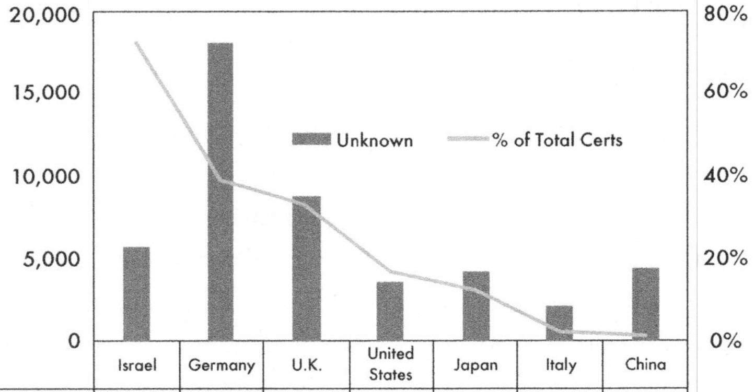

	Israel	Germany	U.K.	United States	Japan	Italy	China
Unknown	5,699	18,086	8,755	5,699	4,099	2,030	4,286
% of Total Certs	72%	38%	33%	16%	12%	2%	1%

Figure 6.9: ISO 9001 Certificates: Unknown Sector by Country

Flaws in the ISO Survey

The ISO survey data was collected from the international CBs that maintain the ISO 9001 certificates. There is no process in place to validate the accuracy of the reporting by the CBs. The ISO recognizes that important CBs in several countries do not participate in the survey. Other CBs are reluctant to report canceled certificates, which would not be helpful to their marketing programs. The inaccurate tabulation and reporting of certified ISO 9001 companies is disappointing (and ironic). Providing a viable customer-feedback process is the cornerstone for companies that maintain a QMS and a requirement for ISO 9001:2015–certified companies. It is also disappointing that the ISO leadership would sponsor the collection of data that is inaccurate and incomplete without conducting analysis and follow-up. The ISO has provided this explanation of the 2018 results:

> The overall total number of valid certificates was lower than in 2017. The reasons for the decrease

are related to the participation and can be categorized into four categories:

- Some large certification bodies reported in past surveys the number of certificates that included the number of sites. In this survey, they have split the number of certificates and the number of sites which led to important reduction[s] in the number of certificates reported.
- The data reported by some large certification bodies fluctuates from one year to another.
- Some providers of data adjusted the way they report the number of sectors covered by their certificates with the separation of the two questions in the 2018 survey questionnaire (they were linked in past surveys which led to some confusion in the case of [multisector] certificates).
- Some certification bodies that are important in some countries did not participate.[1]

For the reasons listed, the ISO decided to publish the 2018 results without the past results. With the adjustments made by some of the providers of data in relation to the number of certificates, sites, and sectors, any comparison of the figures on the number of valid certificates with the past survey would not lead to accurate conclusions. The level of valid certificates in 2018 is a better reflection of the situation in the market. Here are a few observations drawn from my analysis of the 2018 ISO survey:

1 https://www.iso.org/the-iso-survey.html

- Since 2006, the number of US companies certified to ISO 9001 has decreased by 51 percent.
- The number of worldwide companies certified to ISO 9001 dropped by 17 percent from 2017 to 2018.
- The ISO survey database is flawed: survey data from previous years included a mixture of certified sites (i.e., plants) and companies.
- The CBs are not accurately reporting the industry-sector code for companies they certify in several countries.
- On a worldwide basis, nearly an equal number of service-based companies and manufacturing companies are certified to ISO 9001.
- The US has 2,248 service-based companies certified to ISO 9001 (10 percent of total US certificates).
- The industry sector in the United States with the highest number of ISO 9001 certificates is "unknown sector."

The next chapter describes the automotive, aerospace, telecommunications, and medical-device industry certification schemes and the impact they have had on the number of companies certified to ISO 9001.

7

Industry Sector-Specific Standards

The automotive, aerospace, telecommunications, and medical-device industries have developed their own QMSs. Currently, approximately a combined total of 105,000 companies in the world maintain certification to the industry-specific standards TS 16949 (auto), AS 9100 (aero), TL 9000 (telecommunications), and 13485 (medical). Many of the TS, AS, and TL companies also maintain separate ISO 9001 certificates, while medical-device companies may opt to have only a 13485 certificate, which is built along the basis of ISO 9001. I am aware of some AS 9100-certified companies that discontinued their ISO 9001 certifications by convincing their nonaerospace customers that conformance to AS 9100 requirements assured conformance to all the ISO 9001 requirements. Other companies make products for all three of the sectors, plus some other specific certifications, such as Nadcap[1] (for testing, heat treating, and welding) and Underwriters Lab (UL) for electrical.

1 Nadcap was formerly known as NADCAP, for the National Aerospace and Defense Contractors Accreditation Program.

A rubber-products manufacturer I audited several years ago continues to maintain certifications with ISO 9001, ISO 14001, AS 9100, plus UL and Nadcap. The quality manager complained to me that he was being audited almost every month by an outside agency, which was really not an efficient use of his time or the company's resources (Figure 7.1).

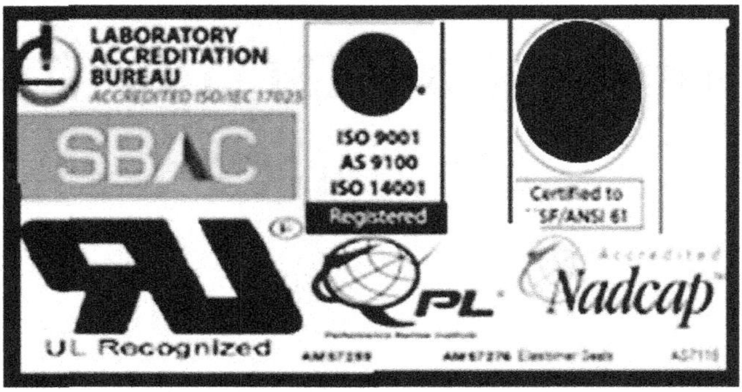

Figure 7.1: Multiple Certificates for Rubber Manufacturer

Automotive: TS 16949. The International Automotive Task Force (IATF) standard IATF 16949:2016 establishes the requirements for a QMS specifically for the automotive sector. The ISO/TS 16949 was originally created in 1999 to harmonize different assessment and certification schemes worldwide in the supply chain for the automotive sector. The primary focus of the IATF 16949 standard is the development of a QMS that provides for continual improvement, emphasizing defect prevention and the reduction of variation and waste in the supply chain. IATF 16949:2016 is an independent QMS standard that is fully aligned with the structure and requirements of ISO 9001:2015. Therefore, the IATF 16949 cannot be implemented as a stand-alone document but must be implemented as a supplement and in conjunction with ISO 9001:2015.[1] Approximately sixty-five thousand companies worldwide are certified to IATF 16949.

Aerospace: AS9100. AS9100 is the international QMS standard for the aviation, space, and defense (AS&D) industry, created by the International Aerospace Quality Group (IAQG); AS9100 Rev D (or AS9100D) from 2016 is the most recent version. The standard provides suppliers with requirements for creating and maintaining a comprehensive quality system for providing safe and reliable products to the AS&D industry, as well as civil and military aviation requirements. "AS9100 certified" means that an organization has met the requirements of AS9100D. AS9100 uses the ISO 9001 standard as a foundation for its requirements and adds certain requirements and notes throughout that are specific to the AS&D industry.[2] Approximately twenty thousand companies worldwide are certified to AS9100.

Telecommunications: TL 9000. The TL 9000 standard, which is the telecom industry's unique extension to ISO 9001:2015, includes supplemental requirements in the following areas: (1) performance measurements based on product reliability; (2) software development and life-cycle management; (3) requirements for specialized service functions, such as installation and engineering; (4) requirements to address communications between telecom network operators and suppliers; (5) reporting of quality measurement data to a central repository. Approximately 1,300 company sites are registered to TL 9000.[3]

Medical Devices: 13485. The standard known as ISO 13485, "Medical Devices—Quality Management Systems—Requirements for Regulatory Purposes," is an internationally agreed-on standard that sets out the requirements for a QMS specific to the

1 https://www.aiag.org/quality/iatf16949/iatf-16949-2016.
2 https://as9100store.com/aerospace-standards.
3 https://www.questforum.org/about-tl9000/tl-9000-certifications/.

medical-devices industry. ISO 13485 is a stand-alone QMS standard derived from the ISO 9000 QMS series. ISO 13485 adapts the ISO 9000 process-based model for a regulated medical-device-manufacturing environment. ISO 13485 certification does not require the organization to also certify to ISO 9001:2015.[1] Currently, approximately twenty thousand companies worldwide are certified to 13485.

Nadcap Certification. Nadcap represents company-level accreditation based on specific processes used by aerospace suppliers, such as chemical processing, coatings, conventional machining, heat treating, nondestructive testing, and welding, among others. Nadcap accreditation is an industry-driven program where highly qualified people who have experience in the aerospace industry conduct the actual process audits, using criteria specific to an actual process. This accreditation is intended for organizations that do business in the aerospace industry. Organizations that wish to become Nadcap accredited will undergo an audit from a Performance Review Institute (PRI) auditor. The auditor will review specific questions related to the process or processes being evaluated. Before attempting Nadcap special-process accreditation, the organization must have an approved QMS such as AS9003 or ISO/IEC17025; if the organization doesn't already have an approved QMS, Nadcap can conduct an AQS audit. Nadcap-certified companies usually also maintain AS 9100 certifications.[2]

---

1 https://www.iso.org/standard/59752.html.
2 https://p-r-i.org/nadcap/.

8

ISO 9001:2015 Interpretation Guidance

In 2014, I decided to go back to my original post-Polaroid partial-retirement plan to provide consulting for companies that required assistance in establishing quality- and environmental-management systems and certification to ISO 9001 and ISO 14001. At this point, with fifteen years' experience auditing hundreds of clients to both standards, I felt I had a solid understanding of ISO 9001 and ISO 14001. I really enjoyed the environmental audits. I could share my working experience related to environmental controls during my Polaroid and Furon career, combined with the knowledge I'd gained from the numerous audits I had conducted, with the environmental-management world. To advertise the services I planned to provide, I created a website, MPD-QE-Consulting.com.

I also decided to write a how-to manual based on my ISO 14001 experiences. After a few months of organizing my thoughts, I contacted Quality Press, the publication arm of the ASQ. Quality Press requested that I send my manuscript to them. After approval by ASQ's peer reviewers, the book, *The ISO 14001:2015 Implementation Handbook*, was published in April 2016. The environmental handbook outlined how an organization could use a management system to both control and improve its environmental performance, linking the management system to the requirements of ISO 14001 to support third-party certification to ISO 14001:2015. Included in the text are best practices as well as common pitfalls and weaknesses I had observed in various organizations during my audits. For those organizations already certified to ISO 14001:2004, the book highlighted the changes required to upgrade to the new 2015 version.

Based on the success of the environmental management system (EMS) book, ASQ asked if I would produce a similar handbook covering implementation of a QMS and providing guidance on certifying to the recently issued ISO 9001:2015. In August 2016, ASQ published *The ISO 9001:2015 Implementation Handbook.* The handbook was structured to guide organizations that were new to ISO 9001 through the process necessary to connect their current practices to the requirements of ISO 9001:2015. For organizations that were already certified to ISO 9001, I provided advice on how to use the upgrade to ISO 9001:2015 as an opportunity to rebuild the QMS into a helpful asset in managing their business. Several readers of the ISO 14001:2015 handbook asked ASQ if a similar handbook could be written to cover the newly issued ISO 45001:2018 standard, related to OH&S management systems. *The ISO 45001:2018 Implementation Handbook* was published by ASQ- Quality Press in July 2018.

More than three thousand copies of my implementation handbooks have been sold by Quality Press and Amazon. ASQ presented me with the Golden Quill award for *The ISO 9001:2015 Implementation Handbook* in 2017. During the preparation of the three books and my many years' auditing experiences, I accumulated a reservoir of best practices in the application of ISO

9001 to a wide variety of companies. The requirements for the design, outsourcing, and servicing processes of ISO 9001:2015 are open to interpretation and are often difficult to apply to organizations of different sizes or complexity. This chapter provides guidance on satisfying the design, outsourcing, and servicing requirements based on my observations in consulting with various organizations. In addition, as noted earlier, ISO 9001:2015 injects risk-based thinking (RBT) or risk management as a key component of QMS certification and as a replacement for preventive action. ISO 9001 is a quality-management support tool. Risk management for an organization is central to the management of the organization's *total business.* The ISO 9001 standard does not provide the structure or auditable requirements to allow assessment of the organization's business risks. The requirements of quality risk management can be efficiently linked to the *change control* requirements of ISO 9001.

Risk-Based Thinking and Change Control

The inclusion of RBT in ISO 9001:2015 has generated a lot of interest. RBT is now a cottage industry, with an extensive gallery of consultants, books, and webinars. The writings I have observed include such profound concepts as "Risk is like swiss cheese" and "RBT is everyone's job." The ISO Auditing Practices Group (APG) uses a person crossing the street (with all the related challenges) in the group's guidance document to explain the RBT process.

Will RBT as a cultural change for organizations be successful in reducing quality or business upsets for organizations of all sizes and complexity? In the past fifty years, there were several quality-improvement initiatives established that attempted to bring about a change in a company's *culture,* or the way company managers and workers interact to improve the company's performance. Quality circles in the 1970s, total quality management (TQM) in the 1980s, and business process reengineering (BPR) in the 1990s were implemented in a wide variety of companies. While each had some successes, very few organizations have maintained these programs as core tools in the operation of their business.

Quality initiatives that *have* been successful over the last several years include Six Sigma and lean manufacturing. These two programs were originally successful and continue to be sustained because, while they do include some level of cultural change or different thought processes for the organization's employees, they are also *results driven.* Top management support is important, although the middle managers and work staff can operate quite independently and effectively once management provides them with a charter for a Six Sigma or lean initiative. I experienced and survived all these initiates in my work life over the same fifty years. The one common denominator in quality or product failures I have experienced or observed is the lack of disciplined change control. I suggest that a company that has a robust and disciplined culture of change control is the ultimate risk-based thinker.

The tragic ignition-switch failure at General Motors is a prime example of an organization's lack of both change control and quality record keeping. The essential problem, according to CNNMoney:

> The cars' ignition switch, where the key is inserted and turned to start the car, could easily be bumped or moved out of the "Run" position into the "Off" or "Accessory" position. When that happens, power braking and steering, as well as airbags, can stop

> working. What most people at GM didn't know was that Delphi, the company that supplied the switch, had redesigned the part in 2006 to make it harder to turn. The problem had been fixed. A GM engineer even signed off on the changes. Unfortunately, GM didn't change the part number of the switch. As a result, manufacturing records didn't indicate that the issue had been resolved.[1]

Several failures are evident in this example. The automotive sector has a requirement for new-parts approval known as the production part approval process (PPAP), which is a generic part-qualification process used to determine if all customer requirements are understood by a supplier and if the process has the potential to make products that meets requirements on a production basis, according to the Automotive Industry Action Group (AIAG) definition. PPAP is a very powerful tool, particularly as it relates to change control and risks in product changes. The process takes time, however, and organizations take shortcuts. A drawing-revision change requires fewer management approval signatures than a new part number and can be implemented quickly. Organizations that use PPAP would be better served to streamline the approval process as opposed to bypassing it!

When the oil rig explosion occurred in the Gulf of Mexico in April 2010, there was much discussion concerning the role of the blowout preventer in mitigating the extent of the spill: Had the risk related to the operation of the blowout preventer been reviewed as thoroughly as needed? Were the maintenance records up to date? When the planned replacement of the blowout preventer was postponed, was the change in procedure properly

1 Peter Valdes-Dapena, "GM Recall: 10 Things You Need to Know," CNNMoney, March 30, 2014, https://money.cnn.com/2014/03/30/autos/general-motors-recall-faq/index.html.

described, approved, and controlled, or was the cost of lost production time the dominant factor?

Other deadly examples where lack of control when making design or production changes include the failures of Firestone tires on Ford Explorers starting in 1998 and the Takata airbags over the last few years. The Takata airbag failures were alleged to have been caused by various changes in work-area conditions (such as moisture) and by operators turning off an "auto-defect reject" system.

I recall that when the Firestone tire failures occurred around the year 2000, there was some pressure to hold the ISO 9001 auditors accountable. A similar concern has now been voiced with ISO 9001 auditors' culpability with the airbag-quality defects, as the supplier maintained a QMS that had been certified by ISO-approved auditors. I am not qualified to express a legal opinion on the subject of culpability, but I will suggest, as a longtime auditor, that we can only assess what the client allows us to see.

Change Control and ISO 9001:2015

How change control (and risk analysis) relates to ISO 9001 requirements can be found in the following processes: (1) process or equipment changes, (2) raw material specification control, (3) document control and review, (4) design, (5) regulatory updates, (6) outsourced processes, (7) planning of internal audits, and (8) effectiveness of corrective action.

One change-control process used by many organizations is the engineering change notice (ECN), which is used to manage how functions or processes interact when changes are made in product, materials, or processes. An almost military-like administration of the ECN process needs to be maintained if an organization is committed to a disciplined change-control culture—and risk protection and mitigation.

Process or Equipment Changes. When production equipment or processes are changed, the implementation plan should include the potential risk for product quality. Testing prior to the release of a product to customers is a common technique to be employed, along with the application of failure mode and effects analysis (FMEA).

Raw Material Specification Control. Any change in materials used in production should be tested before release to customers. The organization should ensure that their suppliers are aware of the need to maintain control and communication of any changes in their specifications or processes.

Document Control and Review. The organization should ensure that any documents used by employees are maintained and controlled to avoid mistakes. Employee instructions should be reviewed at some frequency to ensure that employees are not bypassing operating instructions.

Design and Development. During the design process, a robust verification and validation plan should be employed to eliminate risks related to new designs. The new design process should also include the risk analysis related to the impact the new design process may have on employee safety and the organization's environmental impact, including end-of-life disposal issues.

Regulatory Updates. The organization should maintain a process to be aware of any changes to statutory and regulatory obligations related to its products to eliminate risks related to noncompliance.

Outsourced Processes. Processes that are performed by external parties can create a risk for the organization in meeting its commitments. External-supplier selection should include controls related to the impact the supplier can have on producing acceptable product or services. The inspection of externally supplied products should be based on inspection cost versus risk related to supplier errors.

Effectiveness of Corrective Actions. An important part of the corrective-action process is how effective the correction was to reduce the risk of a recurrence of the same issue. The time and resources allocated to measuring the effectiveness of the correction should be commensurate with the risk of recurrence.

Internal Audits. The internal-audit plan should be based on the impact a process may have on quality performance as well as the history that the particular process has in generating nonconformances. By focusing on the history and impact of each process, the organization can allocate auditing resources to reduce the risk of errors.

Risk Register

One common tool or technique employed in risk analysis and risk management (I don't like the catchphrase of RBT) is the use of a risk register or log. While risk registers include many variants, the key elements are similar to a process FMEA and will usually include (1) a description of the risk, (2) risk type (business, quality, design), (3) likelihood of occurrence, (4) severity, (5) countermeasures, (6) process owner, and (7) status/updates.

The likelihood of occurrence and severity calculations include ranking criteria to help prioritize the countermeasures and actions to mitigate the risks. Depending on the size of the organization and the complexity of its processes, a simplified risk worksheet can be employed. An example of a risk worksheet is illustrated in Table 8.1. Once established, the risk worksheet should be maintained and updated to ensure that the defined risks are mitigated. My clients have found the utilization of a risk worksheet to be a beneficial addition to their QMSs.

Threat	Vulnerability	Current countermeasures	Actions	Responsible
What can happen and consequences	*Weaknesses that make the organization susceptible to the threat*	*Existing components that mitigate or eliminate the vulnerability*	*Control actions?*	*Name*
Loss of raw material (RM) sources	Sole or single-sourced RM; small volume; lack of leverage	R&D surveillance	Develop contingency plan	
Rising energy costs	Heavy dependence on natural gas or electricity	Energy-efficiency team; renewable-energy investments (e.g., solar panels)	Continue efficiency team projects	
Availability of machine operators	Competing jobs in areas	Varies by plant; use of placement agencies	Continue	

Table 8.1: Risk Worksheet

Design and Development

The "Design and Development" clause (8.3) of ISO 9001:2015 contains the following requirement: "The organization shall establish, implement and maintain a design and development process that is appropriate to ensure the subsequent provision of products and services." It is up to the organization to implement a design process that is appropriate to its business model. Clause 8.3 includes the process requirements that an organization might use when designing and developing a highly complex computer hardware system with a design team of a dozen or more members and a project cycle time of more than one year. The majority of organizations certified to ISO 9001 often make minor adjustments or tweaks to existing products with one or two designers, and the project is released to production in a few days or weeks. Organizations with both complex, lengthy projects and quick-turnaround modifications should develop their design-process controls accordingly.

Not explicit in the ISO 9001:2015 requirements are the need to have a *time-bound design plan.* The lack of a planning requirement was also a deficiency in prior revisions of ISO 9001. When I audited an organization's design process in which several designers were working on a variety of projects simultaneously, I was often disappointed to observe the lack of an integrated plan to assist the scheduling of resources. While the required stages of the project may have been defined, the planned completion dates or milestones had not been established. Many organizations use a Gantt chart or similar technique to establish estimated completion dates for each design stage, often working backward from the customer's required completion. By integrating the timelines of all projects, resources can be better allocated. The basic requirements for a design project include:

- statement of work (What is new?)
- time-bound plan
- input requirements/specifications
- outputs/expectations
- design reviews, as appropriate

- verification/validation plan

Dr. Robert G. Cooper, one of the founders of Stage-Gate International, introduced a new product development process, referred to as the trademarked Stage-Gate innovation process, over twenty years ago. Many organizations have used this process or a variation to satisfy the requirements of the ISO 9001 Design clause. A properly designed and implemented stage-gate process will satisfy ISO 9001:2015 clause 8.3 requirements but also add two enhancements while integrating ISO 9001 with the organization's business model. Stage 1 includes a screening of ideas for new developments to analyze the cost benefits of the project, along with potential challenges and competitive factors. The last stage (often stage 8), includes a postlaunch review to measure the project's results and return on investment, as well as lessons learned to apply to future projects. I recommend using some form of the stage-gate process for organizations of any size or complexity (Figure 8.1). The stage-gate model can be used for enhancements/modifications or minor reconfigurations of mature designs, or more complex, long-cycle-time designs. In the case of minor configuration projects with short cycle times, the forms can be designed to allow the consolidation of several steps, with less paperwork and approvals. The related design records from the referenced forms are shown in Table 8.2.

Stage	Form	Contents
Start	Design	Date, customer, contact, unit manager, sales rep., product description, pricing target, other information
1	Analy-sis	Design objectives, business analysis, project leader
2	Plan	Statement of work, initial specifications, project timetable
3	Input	Performance/functional requirements, regulatory impact, previous design review, bill of material
4	Output	Lab prototype, safety/environment impact, pur-chased materials, production sample 1, production/ QA input
5	Results	Repeat production runs, process capability review, verification/customer results, validation at customer, transfer to production
6	Review	Financial analysis, lessons learned

Table 8.2: Design Records

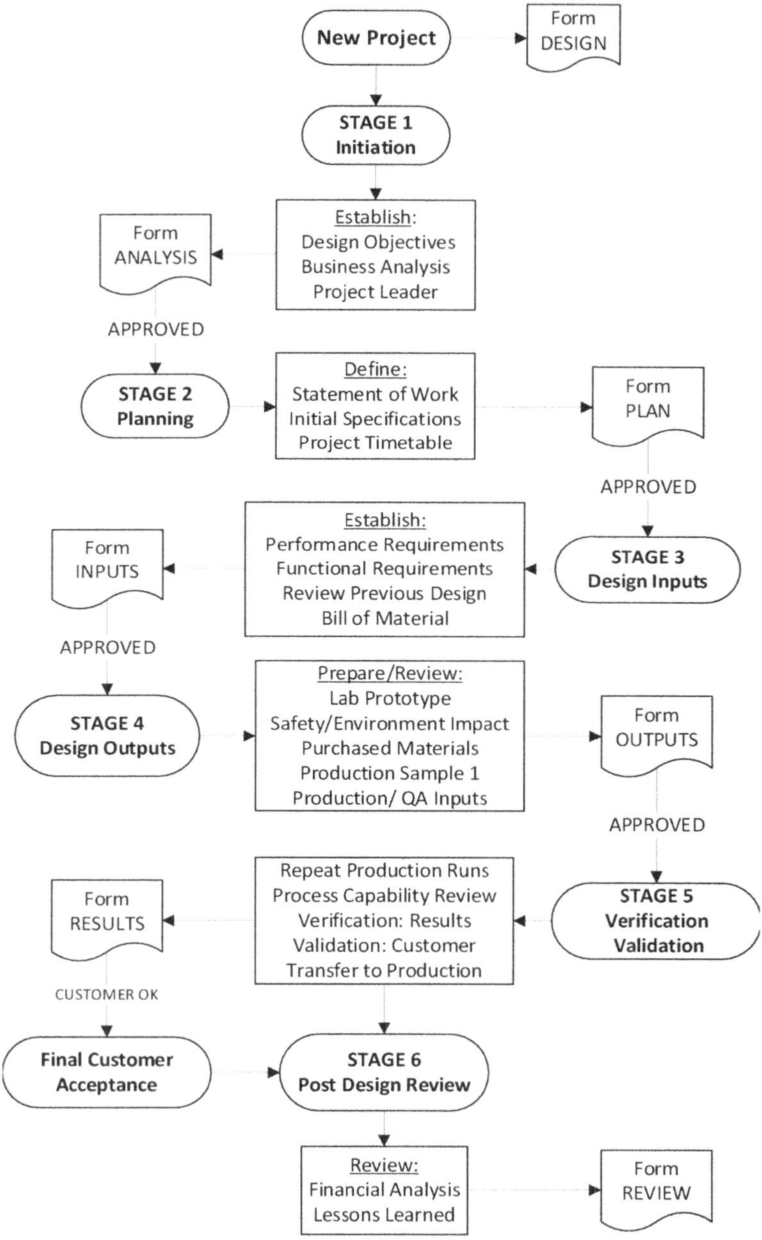

Figure 8.1: Design Flow Chart

For an organization with design projects with various complexities and cycle times, a procedure should accompany the flow chart with definitions and application limitations as described in Table 8.3.

NEW-PRODUCT DEVELOPMENT (NPD) PROCEDURE

Purpose: To establish, implement, and maintain a new product development (NPD) process that is appropriate to the COMPANY'S business and to ensure that the design outputs are consistent with the company's manufacturing capabilities and processes and customer requirements.

Scope: All company locations

Definitions

- **Customer-driven request:** customer inquiry for new design or modification to existing design.
- **Management-driven request:** recognition by the company's business unit managers, sales, or management of the need for a new product or revision to an existing product. These requests may not have direct inquiry from a customer.
- **Verification:** the results meet the requirements of the design-input specifications.
- **Validation:** the new design satisfies the customer's intended use of the material or product.
- The Six Stages of the NPD Process:
- **Stage 1 (Initiation):** project definition, business goals/analysis
- **Stage 2 (Planning):** statement of work, parameters and initial specifications, preliminary timetable
- **Stage 3 (Design input):** performance and functional requirements, regulatory impact, previous design review, bill of material
- **Stage 4 (Design output):** prepare lab prototype, test vs. specifications, safety/environmental impact, bill of material (BOM) review with production, production trials

- **Stage 5 (Verification and validation):** repeat production runs, specification and process-capability review, verification of results, validation for customer application, transfer of specifications and process sheets to production
- **Stage 6 (Results and post design review):** results, financial analysis, lessons learned

Limitations

Projects with cycle time of greater than three months should follow the six stages with appropriate documentation and records. Projects with shorter cycle times should complete stage 1 (Request) and stage 2 (Initiation) using electronic form "New Product Request." Documentation of the other four stages can be retained in laboratory notebooks or similar records. The closing of short-cycle new product developments should be summarized using the Results tab of the "New Product Request" form.

Table 8.3: New Product Development Procedure

A metric or monitoring plan should be established for the design process to measure the effectiveness of the process and to establish a baseline for improvement. A design-oriented organization realizes that the business is at risk in maintaining/increasing sales revenue if new product offerings do not replace products made obsolete by their customers or their market sector. Many organizations I've audited establish a goal for the design group to create incremental annual sales generated by new designs at 15–20 percent of current sales to offset obsolete or poorly selling products. Examples of metrics to track performance in the design process include (1) new sale money for design projects as a percentage of annual sales, (2) percentage of design projects that create sales money, (3) cycle time to complete projects, (4) percentage of design projects completed on time and on budget.

Outsourcing

ISO 9001:2015 does not define outsourced processes. ISO 9001:2008 (clause 4.1) of the general requirements provides the following definition of an outsourced process: "A process that the organization needs for its QMS, and which the organization chooses to have performed by an external party." When an organization purchases materials, products, or services, it is *choosing* to have a process performed by an external party. Outsourcing, in the context of ISO 9001, has been a source of inconsistent interpretation for several years. ISO 9001:2015 does not provide a clear distinction. My personally established definition of outsourcing is as follows: "Materials, products, or services provided for the organization's customers by external providers and shipped directly from the external provider's site to the customer of the organization."

The logic behind this definition centers on *risk* and *control*. For example, an organization engages an external source (supplier) for chrome-plated parts. If the supplier delivers the chrome-plated parts to the organization, then that is *purchasing*. If the supplier delivers the chrome-plated parts directly to the organization's customer, then that is *outsourcing*.

When the parts come back to the organization, they can be inspected by the organization before release to the customer. In the outsourced case, the organization has to rely on the chrome-plating supplier to properly inspect, package, and ship the finished product to the organization's customer. This situation adds the risk that the supplier will not perform the tasks properly. When outsourcing is used in this context, the organization needs to *establish controls* to monitor the supplier's quality. Controls for outsourced sources include a wide range of activities, depending on the quality risk of the outsourced activity. At the high end of risk, the outsourced provider would be required to transmit inspection or functionality results to the organization requesting the work before shipping the product to the organization's customer. Other high-risk-level outsourced processes are managed by having a representative of the organization release the product at the provider's plant. Lower-level controls include certifying the provider via performance history and/or quality audits at the provider's plant.

The outsourcing scenario described above is related to ISO 9001 clause 8 for operations. Other processes may be outsourced related to human resources, information technology, and accounting that require controls appropriate to the activities and risk. When design activities are outsourced, the potential impact on the customer requirements is normally controlled by the hiring organization's approval of the outsourced designer's work.

I recommend that the organization provide the outsourcing definition described previously or provide a definition that is suitable to the organization's business model. The organization should establish and implement appropriate controls to confirm the organization's customer requirements are met.

Service Provisions and ISO 9001:2015

Past revisions of ISO 9001 have struggled to have the product or manufacturing requirements of ISO made applicable to the servicing industry. Over the last fifteen years, I have audited a wide range of organizations in which the *product* they provided to their customers was a *service*. The application of ISO 9001 to these organizations was not so difficult. I believe service-based companies have benefited by employing the discipline of ISO 9001 to their activities. A sample of service companies I have audited includes the following:

- distributor of automotive components in Canada
- distributor of fasteners in South Carolina

- translation and software localization services in Massachusetts
- contract custodial provider at a pharmaceutical plant in Connecticut
- international freight forwarder at Philadelphia International Airport
- engineering/architecture firm in Kentucky
- electrical contractor in Pennsylvania
- project management training firm in Pennsylvania
- ISO training provider in Connecticut
- temporary-help agency in Michigan
- special metal importer in Florida
- engineering/surveying company in Ireland

Of the hundreds of organizations I have audited, I recall these companies most easily—and fondly—perhaps because of the challenge in preparing an audit plan. Many of the locations were also quite interesting. The audit at the architecture company in Kentucky included a site visit at Churchill Downs Racetrack to witness the application of the firm's project planning. The site visit with the electrical contractor in Pennsylvania was at a new school that was being built. During the audit at the ISO training provider in Connecticut, I witnessed an internal auditor-training program being conducted by the organization.

The most interesting audit was with the surveyor in Ireland. The company had been contracted by the Irish government to verify that the plots of land that telephone companies were using to install cell phone towers were properly mapped and owned. The government required the surveyors to hold a third-party quality registration, so ISO 9001 was used. This cell tower land plot was adjacent to an ancient Irish cemetery. I used the surveyor's transit tool to verify his readings.

After a few audits of service-type organizations, I realized I needed to do a better job of understanding the client's processes before preparing the audit plan. I would usually call the client and try to understand their business model before preparing the audit plan. I discovered that there are five general categories of service providers: (1) software or information technology, (2) distributors, (3) sales, (4) engineering services, and (5) other services. The chart below outlines the ISO 9001 clauses that might be relevant to each type of firm. To become certified to ISO 9001:2015, all service organizations need to address the ISO 9001 clauses (Table 8.4).

4: Context of the organization	8.2: Requirements for services
5: Leadership	8.5.1: Control of service provision
6: Planning	8.5.6: Control of changes
7.1.6: Organizational knowledge	8.6: Release of products and services
7.2: Competence	8.7: Control of nonconforming outputs
7.3: Awareness	9: Performance evaluation
7.4: Communication	10: Improvement
7.5: Documented information	

Table 8.4: ISO 9001:2015 Clauses for Service Organizations

Distributor firms that inventory and sell products manufactured by other companies will often have processes that closely resemble a manufacturing site, the major difference being that they do not make products: they sell, purchase, store, and sometimes measure products. Some distributors will provide repair or manufacturing services, so they will have requirements in production and servicing, possibly including the calibration of measuring equipment. Additionally, distributors will have a warehouse. Clause 7.1.3 ("Infrastructure") will need to be considered in relation to transportation and information systems.

Engineering services may provide design activities, such as architecture and building design, so clause 8.3 ("Design and Development Service Requirements") will be applicable. Software developers provide new products, so they also have requirements in clause 8.3; software development is quite different from hardware design, however, and is often a project where only one person provides 90 percent of the work. I would suggest that software developers consider modifying the design stage-gate process to match their business model.

All service-type organizations, when considering registration to ISO 9001:2015, need to carefully review their processes to understand how their activities relate to several matters, including identification/traceability, property belonging to customers, preservation, postdelivery activities, control of changes, release of products/services, and control of nonconforming outputs. While the jargon for a software developer may be different from that of manufacturing organizations, the bugs are a form of nonconforming product or rejected material; their software releases are similar to a hardware maker's revision control of a drawing. The customers provide the software developer with an electronic data file, which is property belonging to the customers as well as intellectual property. To demonstrate how a service organization might approach mapping its process, a flow chart for a software developer is presented in Figure 8.2.

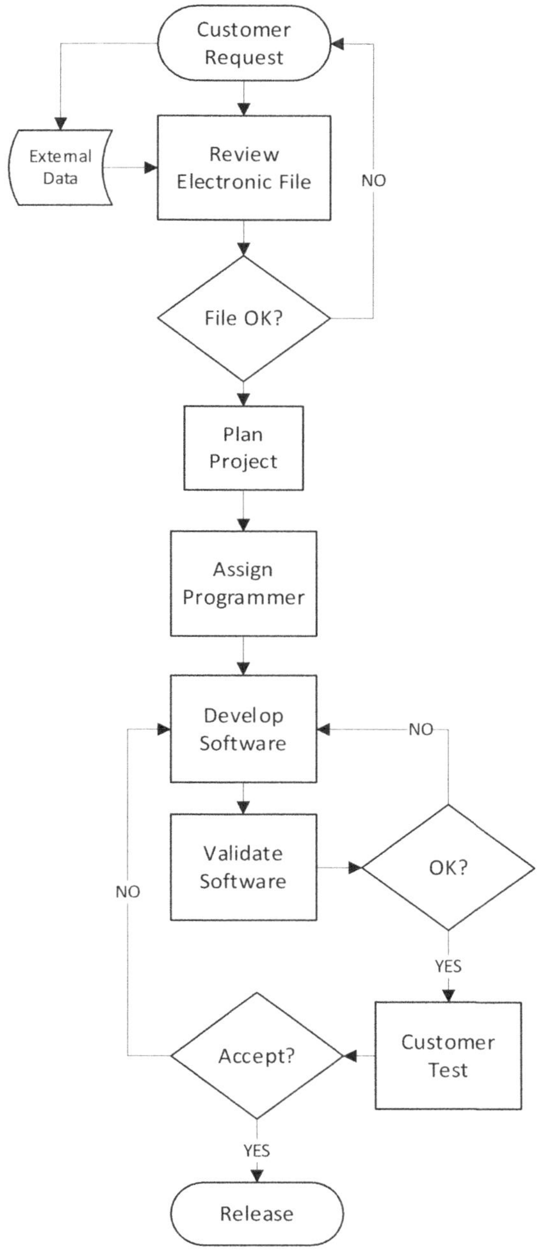

Figure 8.2: Process Chart: Software Developer

For the software developer's processes, the ISO 9001:2015 clauses that may apply are indicated in Table 8.5. For some software developers, the application of clause 8.5.1 ("Control of Service Provision") may be more applicable than 8.3 ("Design and Development"), as some software development can be more process development than product development. The monitoring or measuring requirement related to software development would be the calibration or standardization of the hardware that is used to load and read the new software.

Process	Applicable ISO 9001:2015 Clause	Related Clauses
Customer Request	8.2 Requirements for services	7.5 Documented information
Review Electronic File	8.5.3 Property belonging to customers	7.1.5 Monitoring and measuring
Plan Project	6 Planning	
Assign Programmer	7.1.2 People	7.2 Competence
Develop Software	8.3 Design and development 8.5 Service provision	7.5 Documented information 8.5.2 Identification and traceability 8.5.4 Preservation 8.5.5 Post-delivery activities 8.5.6 Control of changes
Validate Software	8.3 Design and development	7.1.5 Monitoring and measuring
Release Product	8.6 Release of products and services	

Table 8.5: ISO 9001:2015 Clauses for Software Developer

In the sample of my service-related clients, I would say that some of the software developers found effective application of ISO 9001 to their business models, particularly for companies engaged in translation or localization software. Software programmers are a unique breed: creative and often possessive of their individuality. The ISO 9001 discipline has helped some software providers standardize their development processes, which then makes project transfers more efficient.

Service companies should establish a metric to establish a baseline for improvement. Here are some examples:

- distributors/sales: on-time delivery, sales revenue
- software: project cycle time, first-release acceptance rate
- engineering: performance to cost and completion time

While several other service-related organizations have benefited from the application of ISO 9001, those companies that used ISO 9001 certification as a marketing tool to attract new customers were often disappointed with the increased sales results and discontinued their registration after a few years.

9

Causes for the Decline in the Number of ISO 9001-Certified Companies

Several causes explain the major decline in companies choosing to maintain certification to ISO 9001, including the 2008–2009 economic downturn, the cost versus perceived value of certification, and the negative impact of Annex SL on ISO 9001:2015. This chapter goes into more detail on these causes.

The 2008–2009 Economic Downturn

Part of the decline of the number of US companies holding ISO 9001 certification can be attributed to the Great Recession that began in 2008 and the loss of manufacturing companies and jobs in the United States. Cliff Waldman, chief economist for the Manufacturers Alliance for Productivity and Innovation (MAPI), a manufacturing-leadership organization, observed, "US manufacturing got walloped during the Great Recession. It lost 20 percent of its output and 15 percent of its workforce."[1]

1 Laura Putre, "Is the Manufacturing Recovery Real?" *Industry Week*, June 13, 2017, https://www.industryweek.com/leadership/strategic-planning-execution/article/22019638/is-the-manufacturing-recovery-real.

The economic downturn of 2008–2009 resulted in the closure of many ISO 9001–certified plants or companies. While the economy did improve over the next eight years, the number of ISO 9001–certified companies did not increase; in fact, the downward trend continues. Many companies achieve ISO 9001 certification to satisfy their customers' requirements. If certification is voluntary, then the company needs to decide if the cost of obtaining and maintaining an ISO 9001 certificate is worth it.

Cost of Obtaining ISO 9001 Certification

The number of days that the CBs are required to allocate for conducting audits are based on the organization's headcount, with factors related to lack of design function, multiple shifts, and number of employees who perform the same tasks. Currently, ISO 19011:2018 provides guidance on auditing-management systems, including the principles of auditing, managing an audit program, and conducting the management system audits, as well as guidance on the evaluation of competence of individuals involved in the audit process. For ISO 9001:2015, the man-day requirements are defined by the International Accreditation Forum (IAF) in the procedure: "IAF Mandatory Document 'Determination of Audit Time of Quality, Environmental, and Occupational Health & Safety Management Systems.' IAF MD 5:2019" (Table 9.1: ISO 9001:2015 Audit Days).

Annex A Quality Management Systems Table QMS1 (initial audit only)			
Effective No. of Personnel	**Audit Time Stage 1 + Stage 2 (days)**	**Effective No. of Personnel**	**Audit Time Stage 1 + Stage 2 (days)**
1-5	1.5	626-875	12
6-10	2	876-1,175	13
11-15	2.5	1,176-1,550	14
16-25	3	1,551-2,025	15
26-45	4	2,026-2,675	16
46-65	5	2,676-3,450	17
66-85	6	3,451-4,350	18
86-125	7	4,351-5,450	19
126-175	8	5,451-6,800	20
176-275	9	6,801-8,500	21
276-425	10	8,501-10,700	22
426-625	11	>10,700	Follow progression above

Table 9.1: ISO 9001:2015 Audit Days: IAF MD 5:2019[1]

To provide an example of a range of costs for the CB to conduct initial audits, two examples are provided: a twenty-six-employee machine shop and a one-hundred-employee printed-box company.

Machine Shop

From Table 9.1, the CB would need to provide a four-man-days audit for the machine shop. The assigned auditor would spend one man-day conducting the stage 1 "readiness" audit at the organization's site. The stage 2 audit on-site would be three man-

1 https://www.iaf.nu/upFiles/IAF%20MD5%20Issue%204%20Version%202%20 11112019.pdf.

days, but because the organization does not design products—they manufacture products based on their customers' designs—a reduction of 20 percent would be allowed. The total man-days expected would be 80 percent of four days, rounded up to 3.5 man-days. The CBs charge between $1,200 and $1,400 per man-day. Using an average of $1,300, the auditing charge would be $4,550. Additionally, the CBs charge about $500 for administrative costs, for a total CB charge of $5,050. The auditor would likely drive some distance to the organization's site, plus would accrue hotel-meals expenses, adding an additional $750. The total CB charge for initial certification would then be $5,800. From the table, the annual surveillance man-days of 1.5 would be reduced by 20 percent (no design) and rounded to 1.25 days at $1,300/day, which would be $1,625 plus the $500 administrative charge and about $200 for travel and hotel, for a $2,325-per-year surveillance audit for the next two years.

On the third year, a recertification audit would be conducted using two-thirds of the audit days of the initial certification time, or 2.5 man-days or $3,250 plus the $500 administrative fee and about $500 in travel expenses, or $4,250 for recertification.

The twenty-six-person machine shop would spend approximately $14,700 over the initial four years of ISO 9001 certification to the CB. Depending on the machine shop staff's ISO experience, the company would probably engage a consultant to assist in its implementation of a certifiable QMS at a cost of $5,000 to $10,000. Averaging the consultant's charge at $7,500, the twenty-six-employee machine shop would be looking to justify the external expenditure of $22,200 to obtain and maintain an ISO 9001:2015 system, plus the costs generated by incremental activities (company's employees preparing the ISO 9001 documentation, providing employee training and preparing for the audit by the CB's auditor).

Printed-Box Company

A company that designs and prints boxes with a headcount of one hundred employees would be audited for seven days during certification: 2 days for each surveillance audit and 4.5 days during recertification. Using travel assumptions similar to the machine shop and prorated for additional travel expenses would result in the printshop spending about $25,000 with their CB during the four-year certification and maintenance of ISO 9001:2015. The consultant costs would be higher than the smaller machine shop, or $10,000.

The one-hundred-employee printshop would spend approximately $35,000 for external charges during the four years, plus the costs generated by incremental activities of the organization's employees.

Internal Costs of ISO 9001 Certification

Many companies, when working to certify to ISO 9001 to satisfy the requirements of one or more of their customers, understand and reluctantly accept the external costs (CB and consultant expenditures) related to initiating the ISO process. They often complain about the internal resources and time required to support ISO 9001 certification. This is probably the most damning artifact of ISO implementation: *formalizing the organization's quality management system can have a negative impact on the organization's business.*

I recall the closing meeting at a small electronics company several years ago. The company president asked me, "Milt, how much would you estimate it will cost us to become certified to ISO 9001? I want to add some product cost to our customers who have required we become certified." As a third-party auditor, it was not acceptable to discuss such things with the auditee, but I was tempted to counsel the president: his customers might have found the uptick in product cost to be related to the fact that now that the company was ISO certified, their great product quality would be even greater—but a bit more expensive. Not a winning argument with most customers! The general feeling that obtaining and maintaining ISO 9001 certification is a nonvalued financial burden on a company is likely the major reason certified companies do not retain their certification once their customers cease requiring ISO 9001 certification.

During my twenty years' experience with ISO 9001, I would estimate that about 75 percent of the organizations that became certified to ISO during that time did so only because the ISO 9001 certificate was a contractual agreement. The remaining 25 percent were usually self-motivated, or they expected that the ISO certificate would attract new customers. Another reason a company might maintain ISO 9001 certification is the requirement of sector-specific standards, such as TS 16949 (automotive) and AS 9100 (aerospace). These standards require ISO 9001:2015 certification as a foundation for their certification process. The medical-products standard (13485) is not directly linked to ISO 9001:2015, although there are commonalities, and many 13485 organizations hold both certifications.

To provide data as to why ISO 9001–certified companies decide to maintain or drop their ISO 9001 certification, I researched ISO 9001:2015-certified machine shops in Massachusetts. The manufacturing companies in Massachusetts that are grouped under the sector for machine-shop activities are listed at Manta.com. This data source identified 257 companies in Massachusetts with activities directly related to machine-shop processes. Most of these companies maintain a website, which allows access to their ISO status. About twenty-five of the companies are small shops (fewer than five employees) and don't typically maintain ISO 9001 certification.

Figure 9.1 defines the number of machine shops in Massachusetts that maintain ISO 9001 certificates. An analysis of the data indicates that 24 percent of the companies maintain either an ISO 9001:2015 stand-alone certificate or combined ISO 9001:2015 and AS9100 certificates. Twenty-seven companies, or 11 percent of the total sampled, previously held ISO 9001:2008 certificates and did not upgrade to ISO 9001:2015. While companies are sometimes remiss in updating their websites when new standards are issued, I verified that the companies that were still listing ISO 9001:2008 certification had made other entries on their websites in the last few years. Two shops indicated that they maintained 13485 certifications without ISO 9001:2015 certificates.

There are a few takeaways from this sample: 66 percent of the machine shops in Massachusetts that could benefit from ISO 9001 have chosen not to; 11 percent of ISO 9001:2008-certified shops have elected not to upgrade to ISO 9001:2015. Certification to AS9100 required twenty-six of the machine shops to also gain certification to ISO 9001.

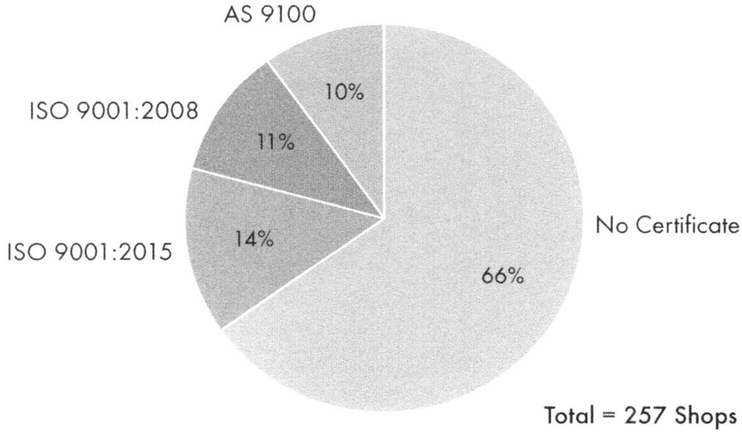

Figure 9.1: ISO 9001-Certified Machine Shops in Massachusetts

To explore why the majority of machine shops in Massachusetts don't see value in obtaining or maintaining ISO 9001 certification, implementing a survey of why the twenty-six shops did not upgrade to ISO 9001:2015—and why 169 shops (66 percent) have not maintained ISO 9001:2015 certification—would be a good approach. I believe these companies would not be open to such a survey from me. The CBs that lost these accounts might know the reasons, but certified organizations often change CBs, which complicates the analysis. As an alternative approach to examining end-user perception of the value of ISO 9001 certification, I sampled a group of ISO 9001–certified companies, as discussed in the following chapter.

10

End-User Survey: ISO 9001:2015

Customer feedback and customer surveys have been a requirement for ISO 9001 for twenty years. In my auditing experiences during that time, I witnessed hundreds of surveys. Most responders represented customers who were either very satisfied or quite disappointed with the supplier's performance. Response rates of under 20 percent were the norm. Often, the most helpful information was the directed comments, where people described specific concerns. The most effective customer-feedback processes I witnessed usually involved contact with a known focus group, where the participants were motivated to respond. For my ISO 9001 end-user survey, I selected two user groups, drawn from the participants of two webinars I conducted for ASQ. After the webinars, forty of the participants requested copies of my presentation and the templates I'd referenced. I emailed those who had quality management responsibilities. Fourteen of the forty quality managers completed the survey (figure 10.1), for a 35 percent response rate. Figures 10.2, 10.3, and 10.4 define each responder's scope of activity, employee headcount, and years certified to ISO 9001. While the sample size was small, the respondents represented a diverse scope of activities and plant locations. The respondents' comments provided some personalized feedback on the pros and cons of the value of ISO 9001 certification.

<table>
<tr><td>Manufacturing / service scope</td><td colspan="2">☐Chemicals ☐Rubber/plastics ☐Metalworking ☐Electrical/optical ☐Machinery ☐Printing ☐Other (specify)</td></tr>
<tr><td>Current certificates</td><td colspan="2">☐ISO: 9001:2015 ☐TS16949 ☐AS9100 ☐13485 ☐No certificate</td></tr>
<tr><td>Number of years of ISO 9001 certification</td><td>☐1-6 ☐7-9 ☐10+</td><td>Headcount
☐< 15 ☐16-50 ☐51-100 ☐101+</td></tr>
<tr><td colspan="3">Registrar performance: ☐Unsatisfactory ☐Satisfactory ☐Excellent</td></tr>
<tr><td colspan="3">Comments:</td></tr>
<tr><td colspan="3">Auditor Performance: ☐Unsatisfactory ☐Satisfactory ☐Excellent</td></tr>
<tr><td colspan="3">Comments:</td></tr>
<tr><td colspan="3">Reason for ISO 9001 Registration: ☐Customer required ☐Corporate required ☐Voluntary
Industry sector required: ☐TS ☐AS ☐13485 ☐Other (specify)</td></tr>
<tr><td colspan="3">Comments:</td></tr>
<tr><td colspan="3">If customer did not require ISO 9001 certificate: ☐Would retain ISO 9001 ☐Would drop certificate</td></tr>
<tr><td colspan="3">Comments:</td></tr>
<tr><td colspan="3">My company's benefits of ISO 9001 certification (check all that apply): ☐Retain customers ☐Attract new customers ☐Improve quality ☐Reduce costs</td></tr>
<tr><td colspan="3">Comments:</td></tr>
<tr><td colspan="3">Negative aspects related to ISO 9001 certification (check all that apply): ☐Cost ☐Time spent for audits ☐Nonvalue-audit findings ☐Inconsistent auditors</td></tr>
<tr><td colspan="3">Comments:</td></tr>
</table>

Figure 10.1: ISO 9001:2015 End-User Survey

Summary

A majority of the end-user responders to the survey indicated that certification to ISO 9001 provided benefits to their organization: 64 percent indicated that ISO 9001 certification helped them retain customers, 57 percent said that certification helped attract new customers, 79 percent indicated that ISO 9001 certification improved the organization's quality performance, and 43 percent felt that certification assisted in cost-saving efforts (Figure 10.5). Half the responders believed that ISO 9001 certification cost too much, 36 percent indicated that the ISO 9001 audits consumed too much time, 36 percent believed that there were inconsistencies among ISO 9001 auditors, and 36 percent felt that the auditors raised nonvalue-added findings (Figure 10.6). The responders had mostly favorable ratings on their CBs (registrars): 29 percent of the responders indicated that their experiences with the CB were "excellent," 64 percent said they were "satisfactory," and 7 percent cited "unacceptable" CB performance (Figure 10.7). The responders had a similar rating for their CB auditors: 50 percent were rated "excellent," 36 percent were "satisfactory," and 14 percent were "unacceptable" (Figure 10.8).

If their customers did not require ISO 9001 certification, 36 percent of the responders would have discontinued their certification to ISO 9001, while 64 percent would have continued with ISO 9001 (Figure 10.9). The comments from the same survey participants were not as favorable as their ratings indicated. Only two of the fourteen comments showed strong favorable expressions of the value of ISO 9001 certification. All responders' comments are included at the end of chapter 10.

Several of the small organizations emphasized that ISO 9001 was designed for large companies; for example, they "should have 'scaled' standards requirements for company size" or the

"ISO 9001 standard [is] too cumbersome for small companies." Another comment described the "game" the organizations and the auditors had learned to play: "Auditors walk a fine line. If they don't give us a good report, we can ask for new auditors. If they don't give us benefit for our money, we can ask for new auditors. They know this and try hard to figure out what upper management really wants."

One responder outlined how independent ISO 9001 certification may no longer be required for organizations that hold sector-specific certification to AS 9100, TS 16949, and TL 9000: "Since AS9100 uses ISO 9001 (word for word) and AS has separated itself from ISO, our customers are satisfied with the more stringent certification of AS9100; a separate cost for ISO 9001 does not benefit our organization or customers as long as we retain our AS9100 certification." This concept has always made sense to me, and may have already accounted for some of the decline in ISO 9001:2015 certifications.

Demographics

The business activity and the number of employees in the responder's company and number of years the company has been ISO 9001 certified.

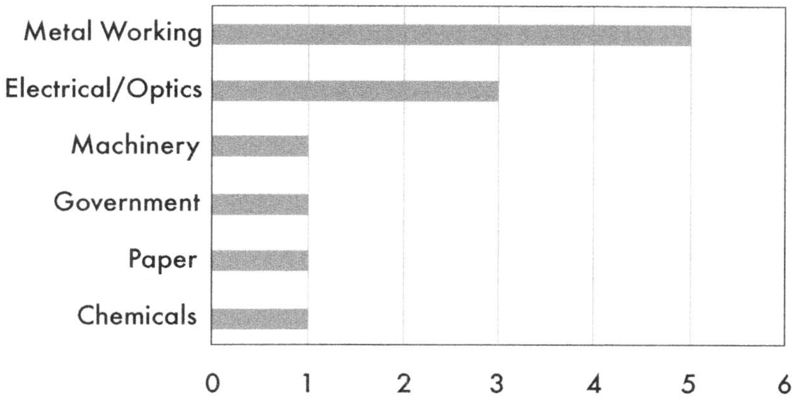

Figure 10.2: Responder's Scope of Activity

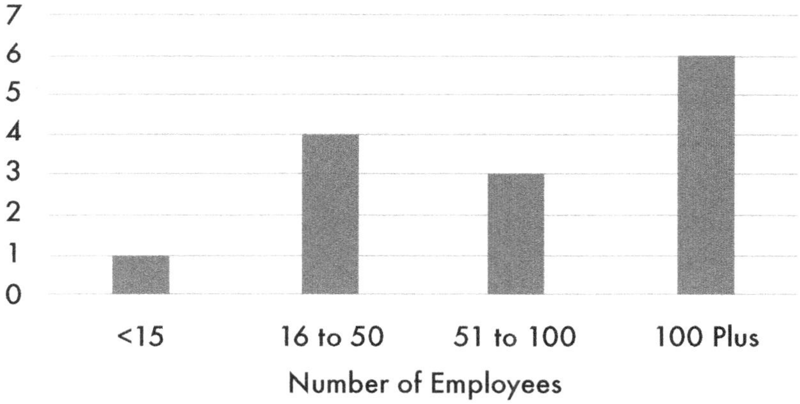

Figure 10.3: Responder Headcount

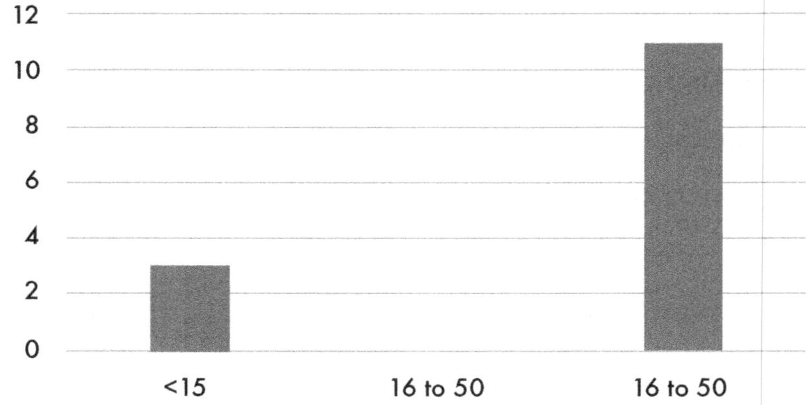

Figure 10.4: Responder's Years of ISO 9001 Certification

Results

The responder's evaluation of the benefits and negative aspects to certification to ISO 9001 and their perception of the performance of their CB and auditor(s). Would the responder maintain an ISO 9001 certificate if their customers did not require ISO 9001 certification?

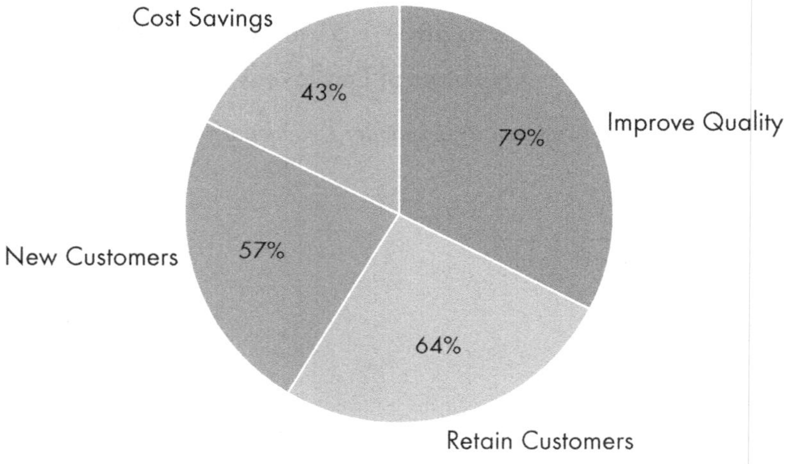

Figure 10.5: Benefits of ISO 9001 Certification

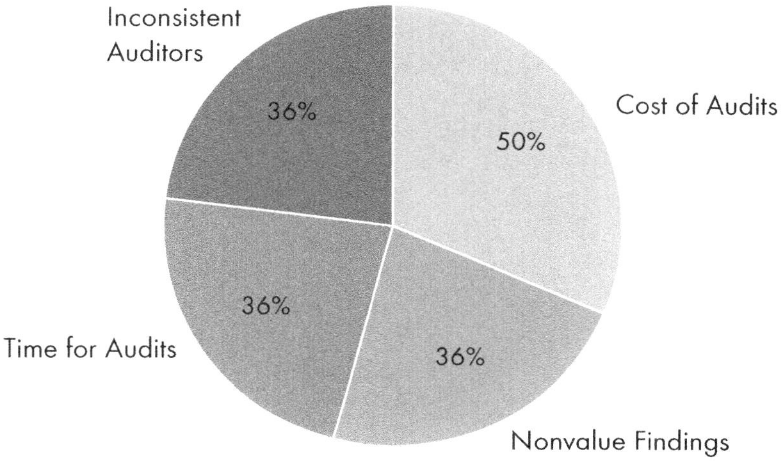

Figure 10.6: Negative Aspects of ISO 9001 Audits

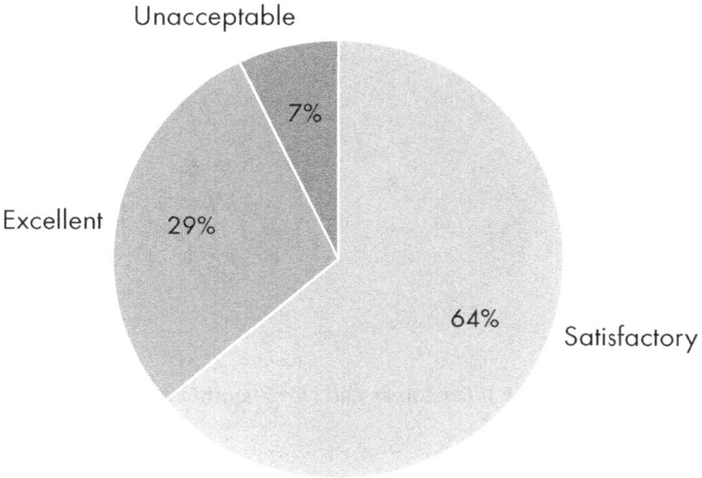

Figure 10.7: CB Performance

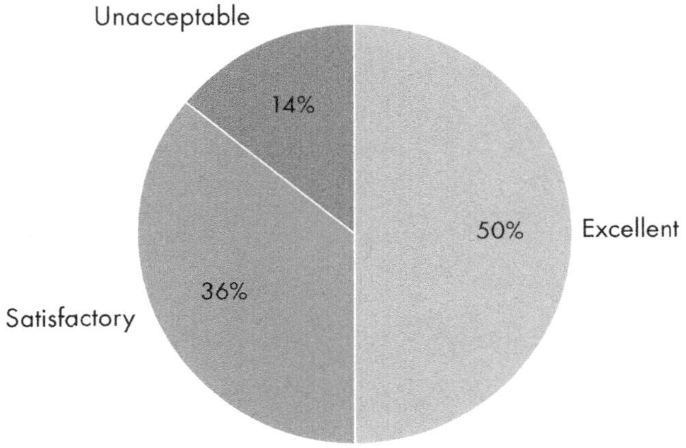

Figure 10.8: Auditor Performance

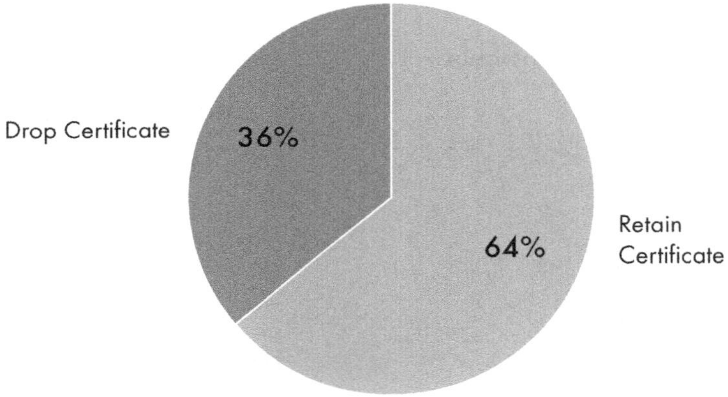

Figure 10.9: If Customer Did Not Require ISO 9001

Responders' Comments

"With the attention devoted to our ISO 9001 certification we have definitely improved both efficiency and costs of our products we produce." (government contractor; 75+ employees)

"The benefits to small companies are not the same as for high production operations. Should have 'scaled' standards requirements for company size." (optical manufacturer; 15+ employees)

"We are in the process of changing our registrar company due to poor performance, meaning that we are not able to pass customer audits for ISO 9001 but we are passing with our registrar. After evaluation of all internal and external reports we came to the conclusion that we need to change the registrar company in order to improve our system." (metal worker; 25+ employees)

"The ISO 9001 standard is too cumbersome for small companies. We only have 8 people. Problems can be found, solutions defined and implemented, with less effort than it takes to document the actions with a paper trail. The management review meeting is not an effective tool for companies with a single manager. We have over 90,000 parts shipped, zero late, zero rejects." (machine shop; 8 employees)

"Yearly surveillance audits and cost, recertification audits after only 3 years and cost, also inconsistent auditors. Audits for the most part are costly and stressful affairs and it doesn't help when the auditors seem to think it's his or her job to make the experience as unpleasant as possible regardless of how long you have been certified." (machine shop; 15+ employees)

"The registrar lead auditors and audit process are excellent. Registrar administration such as scheduling, responding to questions, representative turnover and availability are barely satisfactory. Would probably be able to continue supplying our current customers if we dropped our certificate, but it would probably cause them some concern. The registrar is required to spend more time to audit our operation than is truly needed. This seems to be a wasteful allocation of time for auditors and us." (paper converter; 101+ employees)

"The registrar that we have 'satisfied' our company, in the sense that we know there are 'tougher' registrars in the world, which might be of more added value. We use a registrar who is recognized in our province (it is a state company), but we do know that the manufacturing sector does not constitute the majority of its customer base. The application of the standard, although beneficial, still requires a lot of energy. People (both managers and employees) often have a poor understanding of the standard. The costs of quality are not exploited enough by the standard to show the positive application of the standard." (electrical manufacturer; 101+ employees)

"We rarely talk with the registrar—for scheduling or billing issues, basically. Problems fixed quickly when they occur. No real reason to complain. Auditors walk a fine line. If they don't give us a good report, we can ask for new auditors. If they don't give us benefit for our money, we can ask for new auditors. They know this and try hard to figure out what upper management really wants. In our case, a certificate is more important to upper management than a good QMS. Upper management would not want to spend the money for the certificate if they didn't have to. Without the ISO requirements, upper management would not put any focus on Quality Management. But they would still find someone to blame for poor quality. We once had an auditor fall asleep while interviewing an associate. We had another auditor insist that our process was wrong because we didn't do it the way he had when he was a Quality Manager at another company." (metalworking manufacturer; 100+ employees)

"A handful of customers have made it a purchase order [PO] requirement but the company feels it helps their processes achieve their goals. Without ISO and its guidance, the company's management system would operate in disarray. Continued successes and improved operations can be attributed to having the ISO system. Registrar auditors have been consistent, honest and helpful with their auditing activities." (machine shop; 51+ employees)

"Changed registrar due to poor administration of audit scheduling and accounting. Most auditors were good while others had poor behavior and worse personal hygiene habits. Began with ISO 9001 in 2005 and quickly became involved in AS9100 and ISO 13485 due to changing business needs. IAQG separated from ISO 9001 (resolution 154) and therefore have dropped ISO 9001 certification. Since AS9100 uses ISO 9001 (word for word) and AS has separated itself from ISO, our customers are satisfied with the more stringent certification of AS9100; a separate cost for ISO 9001 does not benefit our organization or customers as long as we retain our AS9100 certification." (metalworking manufacturer, 51+ employees)

"Every time the auditor changes, the style of auditing changes and new observations are identified. It seems the auditor's ego needs to be satisfied. The ISO is integrated into the business processes and we've seen an improvement over the years. Now the structure is in place and this is giving dividends." (chemical manufacturer; 16+ employees)

"Small company—large global registrar. [High-tech] companies require registration. We have a global customer base. If it wasn't required, we probably would not accept the expenses of registration. The benefits to small companies are not the same as for high production operations. Should have 'scaled' standards requirements for company size." (electrical manufacturer; 15+ employees)

"We have had our current registrar for 25 years. We have only had a few auditors that my company was not satisfied with. Some of the added criteria in the 2015 revision are hard for auditors to understand, know what to ask for and if it's in compliance. The language is not easy. Risk based thinking is difficult for some internal auditors; understanding what questions to ask and what evidence to look for." (textile, paper converter; 101+ employees)

11

The Negative Influence of Annex SL

In becoming knowledgeable of the changes in ISO 9001 and ISO 14001 required by the 2015 revisions, written under conformance to Annex SL guidelines, I became aware of the issues created by forcing the environmental management system requirements to align with the QMS requirements. Additionally, when I began work on the *ISO 9001:2015 Implementation Handbook*, I concluded that conformance to Annex SL guidelines served to seriously weaken both MSSs. In describing the implementation plan for readers interested in certifying to ISO 9001:2015, I based my guidance on conforming to the clearly defined requirements of ISO 9001:2008 and overlaying the new requirements of ISO 9001:2015. I believed that ISO 9001:2008, although it had some flaws, was a workable MSS. ISO had evolved over the previous several decades from "document what you do—do what you document" to the 2008 revision, where focusing on the customer and engaging in process management were key requirements. I thought that was a good move. It worked well with many of the customers I audited. I became proficient in assisting companies in preparing process maps and establishing process metrics. The management of successful businesses understood the need to focus on the customer and would establish performance measures to form a baseline for improvement.

Annex SL contains several flaws, in my opinion. By forcing commonality on the quality, environmental, and OH&S clause requirements, the new standards are not as clearly presented as their predecessor standards were. In conforming to Annex SL requirements, the technical committees made changes to ISO 9001:2008, ISO 14001:2004, and BS OHSAS 18001:2007 (Occupational Health & Safety Assessment Series) that weakened the standards without providing value: (1) the new term "documented information" provides unnecessary confusion; (2) the forced alignment of ISO 14001 and ISO 45001 standards with ISO 9001 created non-value-added requirements for those standards; and (3) the commonality of requirements for all standards caused the dilution of several important ISO 9001 quality requirements.

"Documented Information" Terminology Change

Under Annex SL, the new term "documented information" was created to describe what were formerly referred to as documents and records. As described by Annex SL, if the organization needs to prepare information and instructions for describing what needs to be done, it is *maintaining* "documented information." This information had formerly been described in the ISO terminology as procedures, work instructions, standard operating procedures (SOPs), and forms. According to Annex SL, if the organization needs to provide information to validate performance or results, they are *retaining* "documented information," which was formerly defined as records. When the TMB was asked to explain why "documented information" is now used instead of documentation or records, they answered as follows:

> The standard has been updated to reflect current technology. Data, documentation and records are now frequently processed electronically. Therefore, the new term "documented information" has been created to describe and take account of this situation. The term subsumes the previous concepts of documentation, documents, documented procedures and records.[1]

The new MSSs do not require organizations to replace their existing documents and records terminology with "documented information." Annex A1, "Structure and Terminology," of ISO 9001:2015 states that "there is no requirement in this International Standard for its structure and terminology to be applied to the documented information of an organization's quality management system." In my ISO handbooks, I recommend that organizations should avoid disrupting their management system documentation and disregard the new term "documented information." Quality, environmental, and safety records have been an important component of manufacturing and servicing companies for three decades. Any effort to obfuscate or diminish the importance of management system records is not helpful, in my opinion. Quality records are often a key component in resolving product-performance disputes, even lawsuits. Environmental and employee safety records can be instrumental in avoiding financial penalties, and sometimes they can be evidence in criminal cases.

When the technical committees prepared the new standards, they were consistent in using "retaining documented informa-

1 TMB/JTCG Advisory JTCG/TF4/N28, December 3, 2013.

tion" in place of "records." Unfortunately, the requirements for "maintaining documented information" are used rarely in the new standards, particularly in ISO 9001:2015, so the need to document many of the key quality procedures, such as corrective action, nonconforming material, and internal audits, is open to interpretation. The vagueness on what needs to be documented is possibly the most damaging artifact of Annex SL. The clients I have assisted in responding to the new standards have maintained the traditional terminology for documents and records and have successfully upgraded to the 2015 standards for ISO 9001 and ISO 14001. I have also recommended that they document procedures based on the prior standards.

In his book *Understanding the New ISO Management System Requirements*, Dr. David Brewer claims that the reason for changing the terminology was for the ISO to adapt to modern technology. He and the TMB were concerned that if a company in today's modern age used a web page, the web page could contain both records and procedures. The ISO decided to use a single item to cover both documents and records. In my opinion, Annex SL changed the universally accepted, thirty-year-old terminology of documents and records to satisfy the very few companies that might use the internet to manage their "documented information."

The overarching principle in documentation should be to formalize what is needed to ensure that users of the documentation have a source for information and instructions that are accurate and timely, providing consistency in managing the business. Quality, environmental, and safety records provide evidence of conformance to a specification or requirement. Records are the cornerstone to all management systems. Retaining "documented information" as an alternate term for records is both valueless and potentially confusing to employees and management alike.

Forced Alignment of ISO 14001 and ISO 45001 Standards with ISO 9001

The environmental and OH&S management systems have a lot in common. Many organizations maintain an environmental health and safety (EH&S) department. The leader of the EH&S department will often be assigned responsibility to manage the ISO 14001 and ISO 45001 management systems. The two standards contain quite similar clause requirements. The environmental aspects and impacts are analogous to the hazards and risks of the OH&S management system. Both standards have requirements for establishing emergency preparedness and understanding the organization's legal or compliance obligations. The only significant clause difference between the two standards is that ISO 45001 includes "consultation and participation of workers," while ISO 14001 does not. The ISO 9001:2015 standard has many quality requirements that are completely unrelated to managing an environmental or OH&S management system. ISO 9001 has requirements for infrastructure, the environment for the operation of processes, organizational knowledge, the release of products, the control of nonconforming material, customer satisfaction, and design and development that have little commonality with EH&S management.

The TMB of the ISO, in preparing Annex SL, forced commonality for the high-level structure of QMSs with those of environmental and OH&S. This was a fundamental mistake, in my opinion. I am now witnessing where the certification bodies that provide the third-party auditors are misusing Annex SL clause commonality to cause nonvalued audit requirements. At a recent ISO 14001:2015 audit, the auditor required my client to document a description of the organization's processes and interactions. The CB's stage 1 audit ("Readiness Review") instructed the auditor to "describe how the Organization defines the interaction between the processes of the management system." The auditor's finding was: "Need more clarity on the interaction between the processes of EMS."

Many organizations have been certified to ISO 14001 since 1996. They have a strong record of environmental stewardship that won't be improved by the inclusion of a process-interaction flow chart. Satisfying all the clauses of ISO 14001:2015 will provide evidence that the organization is conforming to clause 4.4.

The same auditor at my client's ISO 14001:2015 certification audit included customer satisfaction and customer complaints as areas to be audited. These are clearly ISO 9001:2015 requirements. When questioned, the auditor responded that with the implementation of Annex SL, the ISO 9001, ISO 14001, and ISO 45001 standards all share the common requirement of customer feedback. He couldn't answer how he would link a requirement in ISO 14001:2015 to describe a nonconformance, should the organization not have a process to monitor customer satisfaction.

ISO 9001 Alignment Weakened ISO 14001 in 2004

The attempt to harmonize MSSs using ISO 9001 as the core standard dates back to 2004. The original issue of ISO 14001 in 1996 included this requirement: "Documents are periodically reviewed, revised as necessary and approved for adequacy by authorized personnel." "Periodic" clearly indicated the need for the organization to establish a cycle time for the review of environmental documents. The first revision to ISO 14001 in 2004 removed the word "periodic" to align with ISO 9001:2000 document-control terminology. In environmental (and OH&S) documentation, organizations should establish a *review* process for documentation commensurate with the risks of deviation from employee instructions. The intent of the review process is to ensure that the environmental documentation (work instructions, SOPs) *matches current practices.* Operating personnel often improvise on how they perform a task to either improve their efficiency or save steps. This can't be allowed without management approval. In ISO 9001, the internal audit process may be a suitable method of monitoring *documentation versus practice,* provided audit notes include evidence that work instructions were validated. In most cases, work instructions should be reviewed by the appropriate authority at a defined frequency.

My suggestion would be to have the organization establish a priority list, ranking the *risk level* of work instructions related to the potential for *quality upsets* (or a prior history of error-prone activities). Examples would be the instructions for the final release of sophisticated electronic devices or process conditions for chemical manufacturing. These instructions should be formally reviewed at least annually.

The Technical Management Board Selected the Wrong Problem

I suggest that the TMB for the ISO set out to solve a problem that wasn't the biggest concern for audited organizations and the certifying bodies that provide the auditors for multiple standard audits. I've assisted many organizations in preparing the documentation and implementation plan for multistandard management systems. In the case of an organization maintaining ISO 9001 certification that wanted to add ISO 14001 (or OHSAS 18001), I would utilize the existing appropriate documentation from ISO 9001 (document and records, corrective actions, training, internal audits, and management review) and customize the procedures for ISO 14001, and then add the procedures unique to ISO 14001. Annex SL will not change my approach; in fact, the standards constructed under Annex SL are not as easy to interpret and combine as their predecessors, ISO 9001:2008 and ISO 14001:2004 and BS OHSAS 18001:2007.

I believe the bigger problem in harmonizing the MSSs lies in conducting efficient, organization-friendly audits to multimanagement standards, not the *structure* of the individual standards. When an organization is certified to more than one standard, the certification bodies and organizations have three options available for scheduling the multiple standard audits:

1. Each audit can be conducted independently, at different times by different auditors.

2. The three standards can be audited during the same time frame as a combined audit, with the auditors qualified to audit only one or two of the standards.

3. An integrated audit can be conducted where the audit team is qualified to audit all three standards.

The integrated audit of option 3 is the most efficient in terms of cost and lack of interruption of the organization's day-to-day activities. I have conducted many integrated audits for smaller companies, combining the ISO 9001 and ISO 14001 clauses that had commonality: document and record control, corrective action, training, internal audits, and management review. For larger organizations, where several audit days were required to conduct the audits, the integrated audits were not always possible, as some members of the team were not qualified to audit one of the standards, so the audits would be a combined audit with only a few clauses audited as an integrated process. The implementation of Annex SL did nothing to assist organizations and certifying bodies in moving toward the more efficient integrated audits.

Another challenge for organizations that hold multimanagement systems is being audited separately if the organization holds sector-specific certifications such as automotive TS 16949, telecommunications TL 9000, medical-products ISO 13485, and aerospace AS 9100. The TMB of the ISO would provide valuable assistance to certified organizations with multiple standards and sector-specific schemes if they found a way to alleviate the redundancy that results in separate, independent audits. The problem is not just the third-party cost of the audits; it is the interruption of the organization's daily routine and time required to prepare and respond to the auditors.

12

Flaws in ISO 9001:2015

ISO 9001:2000 and its successor, ISO 9001:2008, were workable QMS standards, in my opinion. The majority of requirements were clearly stated in ISO 9001:2008—experienced third-party auditors could assess an organization's conformance to these requirements with minimal need for interpretation guidance. The ISO 9001:2015 revision contains several flaws. Several of the new requirements in the 2015 version are poorly composed. The clarity of the old requirements that didn't change were diluted. The guidance in the annex to the standard contradicts the text of the body of the standard and is often ambiguous. The issues of concern to me are discussed below.

Lack of Clarity in ISO 9001:2015 Clauses

Several clauses of the ISO 9001:2015 are not clearly expressed. Clause 6.1, "Actions to Address Risks and Opportunities," contains two formatting errors in defining auditable requirements. It is not helpful when a clause references a previous requirement by clause number without including the actual wording of the referenced clause. A clearly defined set of requirements for a clause should be completely defined within the text of the clause.

Additionally, dropping down five levels does not help the reader identify the requirement.

6.1 Actions to Address Risks and Opportunities:
6.1.1 When planning for the quality management system, the organization shall consider the issues referred to in **4.1** and the requirements referred to in **4.2** and determine the risks and opportunities that need to be addressed to: a) give assurance that the quality management system can achieve its intended result(s); b) enhance desirable effects; c) prevent, or reduce, undesired effects; d) achieve improvement 6.1.2 The organization "should" plan: a) actions to address these risks and opportunities; b) how to: 1) integrate and implement the actions into its quality management system processes (see **4.4**); 2) evaluate the effectiveness of these actions Actions taken to address risks and opportunities shall be proportionate to the potential impact on the conformity of products and services

Clause 6.1 includes five levels of indexing (6.1.2.a.1). Why is this necessary? The previous revision of ISO 9001 in 2008 used a maximum of three levels and is much more readable and easy to interpret than the 2015 version. Other clauses in ISO 9001:2015 that provide excessive indexing include clauses 8.1 ("Operational Planning and Control"), 8.5 ("Production and Service Provision"), and 9.2 ("Internal Audit"). Composers of the 2015 standard may defend the need to provide multiple levels of a requirement to assist the clear presentation of gaps or nonconformances in an organization's actions related to the requirement. This concept is contradicted by the majority of the clauses in ISO 9001:2015.

For example, clause 8.5.2 ("Identification and Traceability") contains four requirements ("shalls") but has no indexing. Clause 6.1 would be clearer if expressed as follows:

> **6.1 Actions to Address Risks and Opportunities:**
>
> 6.1.1 When planning for the quality management system, the organization shall consider *the context of the organization and the needs of interested parties.*
>
> The organization shall determine the risks and opportunities that need to be addressed to give assurance that the quality management system can:
>
> a) achieve its intended results
>
> b) enhance desirable effects
>
> c) prevent or reduce undesired effects
>
> d) achieve improvement
>
> 6.1.2 The organization shall plan:
>
> a) actions to address these risks and opportunities
>
> b) how to integrate and implement the actions into its quality management system processes
>
> c) evaluate the effectiveness of these actions.

Lack of Requirement for a Quality Manual

ISO 9001:2015 does not require a quality manual. This is another artifact of Annex SL "documented information" terminology that has the potential to provide more harm than good, in my opinion. How many of the nearly one million organizations currently certified to ISO 9001:2008 will toss out their existing quality manual? Very few, I believe. ISO 9001:2015's Annex A1, "Structure and Terminology," explains why a quality manual may not be necessary or helpful:

> The structure of clauses is intended to provide a coherent presentation of requirements, rather

> than a model for documenting an organization's policies, objectives and processes. The structure and content of documented information related to a quality management system can often be more relevant to its users if it relates to both the processes operated by the organization and information maintained for other purposes.

This guidance is, frankly, incoherent, as is much of the information contained in Annex A of ISO 9001:2015, "Clarification of New Structure, Terminology and Concepts." What should have been communicated in Annex A was this: "Organizations with a quality manual that currently includes paraphrasing of each ISO 9001 clause requirement—going back through several ISO 9001 revisions—should limit the quality manual requirements to those described in ISO 9001:2008." Clause 4.2.2, "Quality Manual," is as follows:

> The organization shall establish and maintain a quality manual that includes:
>
> a. the scope of the quality management system, including details of and justification for any exclusions;
>
> b. the documented procedures established for the quality management system, or reference to them;
>
> c. a description of the interaction between the processes of the quality management system.

ISO 9001:2008 had a clear set of requirements for a quality manual that could be recorded in about a dozen or so pages.

An effective quality manual should be the organization's *policy* statement of how it will manage its QMS in conforming to ISO 9001:2015. The quality manual is a high-level consolidation of the key elements, or *roadmap* of the relevant quality documentation. The recommended contents of a quality manual to include ISO 9001:2015 requirements are as follows:

- a description of the organization's business model, including the context of the organization and the expectations of interested parties;
- the scope (the activities, processes, and buildings and locations) of the QMS;
- a description (and justification) of those ISO 9001:2015 requirements that are not applicable to the QMS, as they do not affect the organization's ability or responsibility to ensure the conformity of its products and services (exclusions);
- the documented procedures established for the QMS, or reference to them;
- a description of the processes and their interaction between the processes of the QMS, including the process used to manage the interactions of the processes;
- the quality policy;
- a description of how the organization has defined responsibilities and authorities to manage the QMS.

Vagueness on What Needs to Be Documented

One of the more damaging changes of Annex SL is the vagueness of what needs to be documented in the management systems. The *documented procedures* of ISO 9001:2008 are now referred to as *maintaining documented information*, and ISO 9001:2008's *records* are described as *retaining documented information*. In ISO 9001:2008, documented procedures were required for six clauses: "Control of Documents," "Control of Records," "Internal Audits," "Control of Nonconforming Product," "Corrective Actions," and "Preventive Actions." When the technical committees prepared the new standards, they were consistent in retaining "documented information" in place of "records." Unfortunately, the requirements for maintaining "documented information" are used rarely in the new standards, particularly in ISO 9001:2015, so the need to document many of the key quality procedures, such as corrective action, nonconforming material, and internal audits, is open to interpretation. A search of ISO 9001:2015 for "maintain documented information" does not indicate that any of the six clauses (preventive actions no longer a requirement) that were required to be documented in ISO 9001:2008 are required to be documented. How many of the nearly one million organizations currently certified to ISO 9001:2008 are going to eliminate their currently documented procedures? How many third-party auditors will allow an organization to manage its QMS without a documented procedure for any of the six baseline processes? Very few, I would estimate. Clause 4.4, "Quality Management System and Its Processes," requires that, "to the extent necessary, the organization shall ... maintain documented information to support the operation of its processes" (4.4.2).

The audited organization and auditor face a quandary: How are the processes evaluated to determine which ones need to be documented? Is the design process more necessary (important) than sales or purchasing? The irony here is that ISO 9001:2008 limited the documented procedures to the six that are required, thus preventing overzealous auditors from forcing excessive documentation on auditees. TC 176 has allowed the "maintenance of documented information" requirements to be arbitrary. The ISO 14001:2015 and ISO 45001:2018 standards were also impaired by the documented information terminology, although many of the clauses in those standards contain the wording, "The organization shall ... maintain and retain documented information."

Clause for Validation of Processes Diluted

ISO 9001:2015 removed a main subclause from ISO 9001:2008, "Validation of Processes" (clause 7.5.2), and moved the requirement into 8.5.1, "Control of Production and Service Provision," as shown here:

> *ISO 9001:2008, 7.5.2: "Validation of Processes for Production and Service Provision"*: "The organization shall validate any processes for production and service provision where the resulting output cannot be verified by subsequent monitoring or measurement and, as a consequence, deficiencies become apparent only after the product is in use or the service has been delivered."
>
> *ISO 9001:2015, 8.5.1-f: "Control of Production and Service Provision"*: "Controlled conditions shall include the validation, and periodic revalidation, of the ability to achieve planned results of the processes for production and service provision, where the resulting output cannot be verified by

> subsequent monitoring or measurement and the implementation of release, delivery and post-delivery activities."

This change is unfortunate, in my opinion, as the validation of processes can be an important component of an organization's quality performance, and it has often been overlooked by both auditors and audited organizations. In ISO 9001:2008, if an organization did not have reason to validate processes, then clause 7.5.2 was excluded with justification for why validation was not required. If an organization can measure its products, either by dimensional, functional, or visual standards, then the organization would not have validation requirements. Examples where the validation of processes is a key component of an organization's quality controls include soldering (where the integrity of the joints can't be verified by measurement), welding, and heat treating or plating.

In these cases, the manufacturer could not ensure that the product met requirements without damaging the product; thus, a process validation is needed to ensure that the product meets the specification. Moving the validation requirement into a lower level of the production and service provision—and removing the important wording, "deficiencies become apparent only after the product is in use or the service has been delivered"—reinforces the idea that validation is only a requirement for a production machine. Validation of the QMS processes has been a requirement since ISO 9001:2000 was issued almost twenty years ago. The change in the 2015 ISO 9001 standard was driven by an attempt to make the quality production and service provision match more closely with the operational planning and control clauses of the environmental and OH&S standards in support of Annex SL.

I often encounter organizations that respond to the validation requirement by including a comment in their documentation—"We will validate any process where deficiencies become apparent only after the product is delivered"—without defining what the processes are, or how they are validated. If organizations measure or monitor all their products before their release to customers, they should declare subclause f of 8.5.1 as not being a requirement.

Lack of Clarity: The Monitoring or Measuring of Processes Requirement

Another unfortunate change was the elimination of the clause relating to the *monitoring and measurement of processes.* This requirement was initiated with the 2000 revision of ISO 9001 and was a major step in causing certified organizations to view their QMS as a series of *processes* that should be monitored or measured to set a baseline for improvement. ISO 9001:2008 clause 8.2.3, "Monitoring and Measurement of Processes," is as follows:

> The organization shall apply suitable methods for monitoring and, where applicable, measurement of the quality management system processes. These methods shall demonstrate the ability of the processes to achieve planned results. When planned results are not achieved, correction and corrective action shall be taken, as appropriate.

Organizations and third-party auditors generally interpreted this clause as requiring organizations to establish metrics (and goals) for their core processes: sales, design, purchasing, production, or service. Support processes should, at a minimum, be monitored as part of the organization's internal audit process. If the organization determines that a metric is not practical or useful for a core process, then that process would also be monitored by the internal audit process. To align with Annex SL guidelines and to have all management systems include the same ten high-level clauses, the writers moved the monitoring requirement into clause 9, "Performance Evaluation." Clause 9.1, "Monitoring, Measurement, Analysis and Evaluation," states that "the organization shall evaluate the performance and the effectiveness of the quality management system."

ISO 9001:2015 has two additional performance evaluation clauses that indicate that the monitoring process requirements are similar to ISO 9001:2008. Clause 4.4, "Quality Management System and Its Processes," includes 4.4.1c: "Determine and apply the criteria and methods (including monitoring, measurements and related performance indicators) needed to ensure the effective operation and control of these processes"; and 4.4.1g: "Evaluate these processes and implement any changes needed to ensure that these processes achieve their intended results."

By combining the requirements of process performance and the monitoring of processes in two disconnected, vaguely defined clauses, the writers of ISO 9001:2015 have done the ISO 9001–certified organizations a disservice. The process monitoring and measuring of the single clause 8.2.3 of ISO 9001:2008 was consistent with the "Plan-Do-Check-Act" concept espoused as the cornerstone of all MSSs. I can't understand the justification for obfuscating this requirement by TC 176, who revised the monitoring and measuring of processes in the 2015 version of ISO 9001. It is particularly disappointing since the spirit of the 2015 standard is to promote the integration of ISO 9001 into the organization's business. Well-run businesses know all about setting hard metrics and will continue to establish metrics and goals in support of their improvement activities.

The Annex to ISO 9001:2015 Allows the Avoidance of Risk-Based Thinking

I believe the addition of risk analysis (and the deletion of preventive action) is a good addition to ISO 9001:2015. Unfortunately, in the annex to ISO 9001:2015, TC 176 made the risk analysis *voluntary*. Annex A, "Clarification of New Structure, Terminology," attempts to clarify RBT, as follows:

> Although (6.1) specifies that the organization shall plan actions to address risks, there is no requirement for formal methods for risk management or a documented risk management process. Organizations can decide whether or not to develop a more extensive risk management methodology than is required by this International Standard, e.g., through the application of other guidance or standards. Not all the processes of a quality management system represent the same level of risk in terms of the organization's ability to meet its objectives, and the effects of uncertainty are not the same for all organizations. Under the requirements of 6.1 the organization is responsible for its application of riskbased thinking and the actions it takes to address risk, including whether or not to retain documented information as evidence of its determination of risks.

Most third-party ISO 9001 auditors will expect some form of documentation to indicate that RBT is part of the organization's QMS. One guideline I use in training internal auditors is that "if it is not documented, it didn't happen." How an auditor can determine if an organization is applying riskbased thinking, without objective evidence (documentation), is contrary to the concept of auditing. From the genesis of ISO 9001 in 1987, anecdotal or verbal evidence has never been acceptable verification for satisfying a requirement.

Annex A, "Clarification of New Structure, Terminology," provides confusion, not clarification. TC 176 should modify this double-speak and clarify what is expected of organizations related to risk analysis. The annexes in ISO 14001:2015 and ISO 45001:2018 provide clarity on how an organization may address risks based on the "context on which the organization operates" (ISO 14001) or the "hazards associated with the activities of the organization" (ISO 45001). This is useful advice, as it allows the organization latitude in establishing risk analysis consistent with its business model.

Lack of Definitions in the ISO 9001:2015 Standard

Annex SL requires the MSSs to use identical definitions for such terms as *organization*, *interested party*, *policy*, *objective*, *competence*, and *conformity*. ISO 9001:2015, ISO 14001:2015, and ISO 45001:2018 approach the inclusions of definitions differently. The quality management standard, ISO 9001:2015, does not include the definitions that are key to users of the standard. To obtain the definitions, the user of ISO 9001:2015 is required to purchase ISO 9000:2015 from the ISO, ANSI, or ASQ for $176 to $209. The ISO 14001:2015 and ISO 45001:2018 standards include the definitions in the standards. Someone who seeks to obtain the QMS standard and related definitions has to spend twice as much as someone seeking the same information for the environmental or OH&S standards. Annex SL allows TC 176 to include definitions in a separate document without explaining *why* definitions—important to understanding the text of the requirements—are not located directly in the standard.

Inclusion of Prescriptive "Notes" in the Standards

Dr. David Brewer, author of *Understanding the New ISO Management System Requirements* (and a sponsor of Annex SL), describes the "notes" added to a management system as follows:

> A note in an ISO management system is intended to assist readers in understanding the requirement. It does not modify the requirement or imply that a particular way of meeting the requirement is itself a requirement. A sure test of one's understanding of the note is that the requirement should not change if the note is ignored.

Three of the standards released after Annex SL was issued provide notes that are a collection of information—some useless and some prescriptive and problematic. As Dr. Brewer also explains, a "note" included at the end of an ISO clause should not provide the organization with advice or options on how to satisfy the requirement, or what a third-party auditor can accept as conformance to the requirement. Some examples of misapplied notes found in ISO 9001:2015 follow:

ISO 9001:2015 Clause 7.1.4, "Environment for the Operation of Processes"
Note: A suitable environment *can be* a combination of human and physical factors, such as:
a) social (e.g., nondiscriminatory, calm, nonconfrontational); b) psychological (e.g., stress-reduction, burnout prevention, emotional protection); c) physical (e.g., temperature, heat, humidity, light, airflow, hygiene, noise)
These factors can differ substantially, depending on the products and services provided.
The problem: The intent of clause 7.1.4 is to cover work-environment issues such as work area temperature and humidity and electrostatic conditions and dirt that can cause nonconforming product. Many products need to be produced (and measured) in a controlled temperature/humidity environment. A third-party auditor is not qualified to make judgments on the social, psychological, or physical condition of workers related to the quality of the product they produce.

ISO 9001:2015 Clause 8.2.3, "Review of the Requirements for Products and Services"
Note: In some situations, such as internet sales, a formal review is impractical for each order. Instead, the review *can cover* relevant product information such as catalogs. **The problem:** This note is prescriptive—but also not helpful. I have audited the catalogs of many organizations; they are often poorly maintained.

ISO 9001:2015 Clause 9.1.3, "Analysis and Evaluation"
Note: Methods to analyze data *can include* statistical techniques. **The problem**: Quite a revelation in 2015! Organizations had been employing statistical techniques for over fifty years to analyze data.

These examples of notes provide much more than "guidance in understanding or clarifying the associated requirement," as defined by the authors of Annex SL. While US TAG 176 may argue that the notes aren't part of the requirements—and are not auditable—the obvious question is: If no one *can* or *will* use these purported clarifications, then why are the prescriptive notes included in the standards?

The approval of the problematic notes in ISO 9001:2015 by the US TAGs concerns me. In the past several years, I have communicated with both TC 176 and the US TAG 176 related to my concerns with ISO 9001. I have been disappointed with the lack of openness of the ISO 9001 technical committees in entertaining suggestions from individuals outside the ISO committees. I believe TC 176 lacks the transparency required by Annex SL, as described in the next chapter.

13

Lack of Transparency: TC 176

Annex SL requires the TCs preparing the MSSs to be *transparent* during preparation and implementation of the MSS. According to Annex SL, section 8.2.4, "Transparency of the MSS Development Process":

> MSS have a broader scope than most other types of standard. They cover a large field of human endeavour and have an impact on a wide range of user interests.
>
> Committees preparing MSS should accordingly adopt a highly transparent approach to the development of the standards, ensuring that possibilities for participation in the process of developing standards are clearly identified, and the development processes being used are understood by all parties.
>
> Committees should provide information on progress throughout the life cycle of the project, including:

> - the status of the project to date (including items under discussion),
> - contact points for further information,
> - communiqués and press releases on plenary meetings, and
> - regular listings of frequently asked questions and answers.
>
> Maximum use should be made of the resources of the ISO Central Secretariat to facilitate the transparency of the project, and the committee should, in addition, consider the establishment of a dedicated open-access website.

According to SL 8.2.5, "Process for Interpretation of a Standard":

> The committee should establish a process to handle interpretation questions related to their standards from the users, and should make the resulting interpretations available to others in an expedient manner. Such a mechanism can effectively address possible misconceptions at an early stage and identify issues that may require improved wording of the standard during the next revision cycle.

The process to have an implementation question answered related to ISO 9001:2015 via TC 176 is clumsy at best, and it appears to discourage questions. I contacted standards@asq.org with an interpretation question on ISO 9001:2015. I was directed to US TAG 176 SOP SC2-TG22-001 revision 1, called "US Guidance for Handling Requests for Interpretation of the Requirements of ISO 9001."[1] The procedure includes a form (not user friendly) to be filled out and forwarded to standards@asq.org. The only acceptable questions are those that can be answered with yes or no. The procedure provides two examples of acceptable questions—and five guaranteed rejections. An example of a US TAG 176 reject is, "Can a consultant audit a company they have consulted with? Are they being impartial?" This question is unacceptable because "this question actually asks two (2) separate questions, both of which can be answered with a 'Yes' or 'No' [e.g., is a yes answer applicable to the first question, the second question, or both?]. Requestor should submit two (2) separate requests, one for each question."

Of the three standards, only TC 207, which prepared the environmental standard ISO 14001:2015, followed the Annex SL requirements for transparency, as indicated by my research. I had a question about interpreting a clause of ISO 14001:2015, so I contacted the secretary of ISO/TC 207/SC 1, who directed me to the committee's procedure 1n1358, "Revised SC1 Interpretation Procedure."[2] TC 207 also maintains a list of ISO 14001 implementation FAQs received from various sources. The committee that prepared ISO 45001:2018 was project committee (PC) 283. (British Standards) BS OHSAS 18001:2007 was not an international standard; it was prepared and administered under the British Standards Institute (BSI). PC 283 was recently approved as a TC 283. I contacted the committee manager for ISO/TC 283 with a question about ISO 45001:2018, and the manager promptly provided an answer. The committee manager also indicated that TC

1 http://users.neo.registeredsite.com/9/2/0/18247029/assets/WG22_Interpretation_Guidelines_SC2-TG22-001_v012717c.pdf.

2 http://users.neo.registeredsite.com/9/2/0/18247029/assets/1n1358_Revised_SC1_Interpretation_procedure_TC_207_ems.pdf.

283 will be forming a FAQ compilation—a good start for a new committee that's administering a new standard.

Both TC 207 and TC 283 allow US users of their standards to request an answer to implementation questions by emailing standards@asq.org, with your name, address, phone, email, and organization or business. Members from other countries can contact their TC's country representatives. TC 176 should review the communication process used by TC 207 and establish a similar approach to assisting interested parties with interpretation issues. The ISO 14001 FAQ website can be very helpful to certified organizations and auditors related to understanding concerns related to interpretations of the environmental MSS. I did not find the clarifications listed on the TC 207 FAQ list to be consulting related. The TC 207 website[1] advises that "it is important to note that Interpretations do not change the requirements in ISO 14001:2015, but are intended to give users a better understanding on ISO 14001:2015. Please note that some [National Board Members] NBM use the term 'clarification of intent' instead of the term 'interpretation.'"

I have always found less need for interpretation guidance with the environmental MSS than the QMS standard. With ISO 14001, the definitions related to the standard are included in the standard, and the annex to ISO 14001, "Guidance on the Use of This International Standard," has clarifying, helpful information. ISO 9001:2015 users have to purchase an additional document, ISO 9000:2015, "Fundamentals and Vocabulary," which provides definitions and guidance. ISO 9001:2015, Annex A, "Clarification of New Structure, Terminology and Concepts," provides confusing guidance and contradicts some requirements of the ISO 9001:2015 standard.

---

1 https://committee.iso.org/sites/tc207sc1/home/projects/published/iso-14001---environmental-manage/iso-14001-interpretation.html.

I believe if the US TAG 176 membership had been more open to inputs from experienced quality professionals outside their committee, the many flaws in the 2015 revision to ISO 9001 described in *The Rise and Impending Demise of ISO 9001* could have been avoided. I have tracked the membership of US TAG 176 for several years. I am concerned that quality consultants and registrars/training companies that get paid to provide services to organizations certified to ISO 9001 may have more influence on the standards produced and approved by US TAG 176 than the certified companies that pay for the services. The next chapter describes the various groups (interested parties) that make up the 2020 roster of US TAG 176.

14

US TAG 176 Membership

The US TAG 176 is responsible for reviewing and approving the QMS standard ISO 9001. In the last ten years or so, there have been concerns that the US TAG 176 was dominated and controlled by consultants who had the potential for financial gain based on their activities within the TAG. Several current and former US TAG 176 members have written guidance books to assist organizations in addressing the requirements of ISO 9001 at the 2000, 2008, and 2015 revisions. Former and current US TAG 176 members have provided paid training and webinars featuring ISO 9001 topics and guidance on interpretation of the ISO 9001 standards. While the TAG members should be allowed to share the expertise they've gained during attendance at TAG meetings, does US TAG leadership approve of TAG members producing books and webinars while the ISO 9001 standards are in the draft phases, prior to official release as the international ISO 9001 standard? To review the current makeup of US TAG 176, I contacted the administrative person at standards@asq.org, who provided me with a copy of the US TAG 176 (2020) roster.[1]

1 http://users.neo.registeredsite.com/9/2/0/18247029/assets/2020-01_TAG_176_Roster_redacted.pdf.

US TAG 176 provides a disclaimer: "You must not use this data for commercial purposes except with the specific consent of the person concerned." I have redacted the TAG 176 member names from the link and have included only each member's company identification. US TAG 176 currently includes 138 members, with 134 listed as participants; four members are noted as "honorary," possibly indicating their status as nonvoting advisers. The members are classified as company (63), consultant (48), individual (14), organization (10), and government (3). Figure 14.1 charts the percentages of each category.

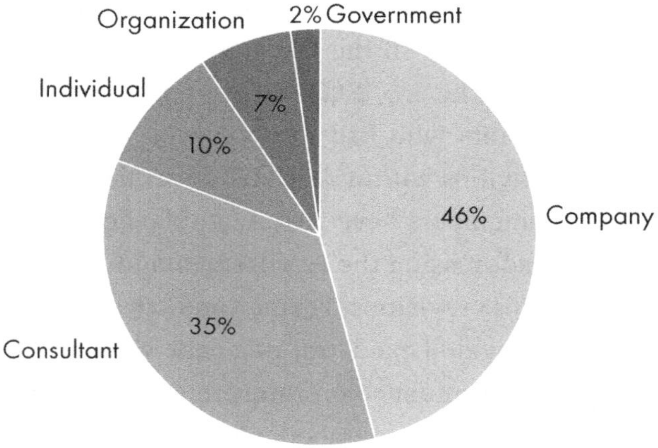

Figure 14.1: US TAG 176 Members, 2020

The TAGs are obliged by ANSI to manage their membership where no single interest category represents a majority of the membership (or, for safety-related groups, no more than a third of the membership), so as not to induce bias in the input or voting on the standard's content. ANSI provides oversight to the US TAGs and has a procedure to define the balance of membership for the TAGs. "ANSI Essential Requirements; Due Process Requirements for American National Standards, January 2019 Edition" specifies in section 2.3, "Balance":

> Historically the criteria for balance are that (a) no single interest category constitutes more than one-third of the membership of a consensus body dealing with safety-related standards or (b) no single interest category constitutes a majority of the membership of a consensus body dealing with other than safety-related standards. The interest categories appropriate to the development of consensus in any given standards activity are a function of the nature of the standards being developed. Interest categories shall be discretely defined, cover all materially affected parties and differentiate each category from the other categories. Such definitions shall be available upon request. In defining the interest categories appropriate to a standards activity, consideration shall be given to at least the following: (a) producer; (b) user; (c) general interest. Where appropriate, additional interest categories should be considered.

The ANSI requirement for membership restrictions for safety-related standards developers clearly defines a maximum percentage for an interest category as one-third of the group's total. Nonsafety standard-developer groups should not allow a single interest to represent a majority of the group's makeup. US TAG 176 appears to be in conformance with ANSI requirements, as no single interest category exceeds 50 percent of the TAG's membership. As Figure 14.1 indicates, the single interest group "company" represents 46 percent of the TAG's membership. In reviewing the US TAG 2020 roster, however, there are inconsistencies in classifying several individuals and the companies they represent. One registrar (CB) is listed as an "organization," five registrars are listed as "companies," and three service providers are also listed as "companies." Other registrars and consultants are properly listed in the "consulting" category. While registrars and service providers are *companies*, for the balance calculation their activities should align with consulting, as they get *paid* for their services. The "companies" (the ISO 9001 end users) on US TAG 176 *pay* the registrars for their services. Subtracting the nine participants from the "company" total and adding the nine to the "consulting" category results in the corrected Figure 14.2.

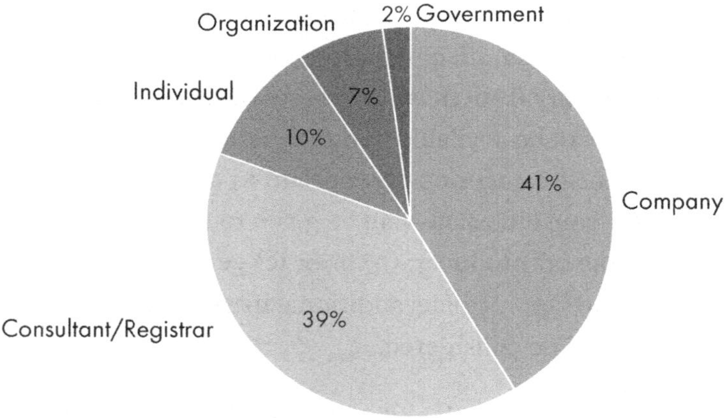

Figure 14.2: US TAG 176 Members—Corrected

The US TAG 176 membership in 2020, while technically in conformance to ANSI TAG membership requirements, appears overly represented by consultants and registration companies that earn their income based on activities related to the content of the ISO 9001 MSS they help prepare and approve. Ideally, a balanced TAG membership would have an equal one-third category allocation based on ANSI guidance terms: producer, user, and general interest. Described in business categories, the three groupings would be: *spend* money on ISO 9001 certification activities, *earn* money from ISO 9001 certification activities, and *neutral* spending related to ISO 9001 certification. The current US TAG 176 appears biased to the "earners," as depicted in Figure 14.3. Of the twelve members listed as "individuals," without company affiliation, at least five could be identified as ISO auditors, consultants, or training providers. Not having precise information on these members, I have allowed the five individuals to be placed in the neutral money category. If these five individuals were instead included in the "earners" category, then the current US TAG 176 roster would have 45 percent of its members earning income from their association with ISO 9001.[1]

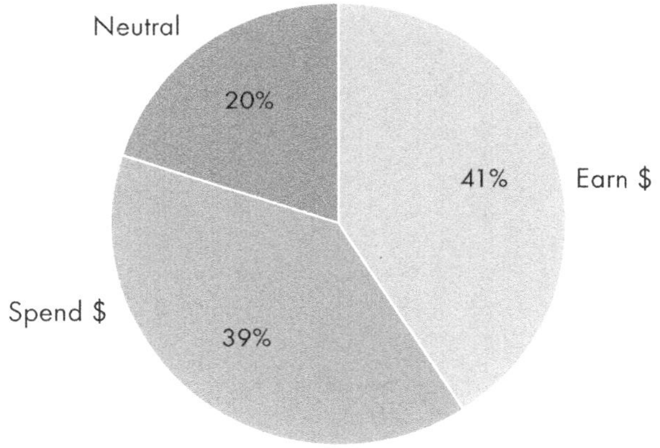

Figure 14.3: US TAG 176: Earn vs. Spend Groups

I recommend that the US TAG 176 review its membership roster to accurately define the interest categories the individu-

1 *ANSI Essential Requirements: Due process requirements for American National Standards*, American National Standards Institute, January 2019, https://share.ansi.org/Shared%20Documents/Standards%20Activities/American%20National%20Standards/Procedures,%20Guides,%20and%20Forms/2019_ANSI_Essential_Requirements.pdf.

al members represent. US TAG 176 should consider following ANSI membership-allocation requirements for safety-related standards developers by limiting single-interest-categories representation at a maximum of one-third of the membership.

15

Corrective Actions: ISO Management System Processes

The creation of Annex SL resulted in a weakened QMS standard, ISO 9001:2015. Starting in 2006, and prior to the issuing of Annex SL, organizations were starting to discontinue their ISO 9001 certification. The issuing of ISO 9001:2015 accelerated the decline of ISO 9001-certified organizations, particularly in the United States. ISO 9001-certified organizations will be irrelevant in the next few years if the leadership of the ISO does not recognize the problem and establish a corrective action plan. The defects related to ISO 9001:2015 processes, the responsible ISO groups, and the corrective actions required are summarized in Table 15.1.

Process	Defect	Responsible Party	Corrective Action
Development of international standards	Annex SL guidance caused ISO 9001 MSS to be weakened	TMB	Rewrite Annex SL
Preparation of ISO 9001	ISO 9001:2015 is a flawed MSS	TC 176	Reissue ISO 9001
Oversight of US TAG 176	ISO 9001:2015 was released in the US containing errors and unclear requirements	ANSI	Approve the US TAG products sold by ANSI
The ISO Survey of Management System Standard Certifications	The certification bodies (CBs) are not accurately reporting the number of worldwide ISO 9001 certificates	Accreditation bodies (ABs)	ABs hold the CBs accountable for accurate reporting
Man-days required for ISO 9001 third-party audits	Organizations do not perceive value in return for costs of third-party audits	IAF	Quantify QMS risk levels appropriate to the organization's processes

Table 15.1: Summary of Corrective Actions

Development of International Standards

As noted earlier, the TMB is responsible for the directives, which are essentially the rules for the development of international standards (ISs), and the TMB deals with all matters of strategic planning, coordination, performance, and monitoring of technical committee activities.[1] The TMB prepared and released Annex SL in 2012. By forcing commonality on the quality, environmental, and OH&S clause requirements, the new standards are not as clearly presented as their predecessor standards were. TC 176 made changes to ISO 9001:2008 that weakened the ISO 9001 standard without providing value. The new terminology of "documented information" provides unnecessary confusion; the forced alignment of ISO 9001 with the ISO 14001 and ISO 45001 standards created nonvalue-added requirements for all three standards; the commonality of requirements for all standards caused the dilution of several important ISO 9001 quality requirements.

Annex SL should be modified or rescinded, and the TCs should be allowed to use the former MMSs guidance when preparing MMSs.

Preparation of ISO 9001

TC 176 did not conform to all the Annex SL requirements when preparing ISO 9001:2015. As described in Chapter 12, "Flaws in ISO 9001:2015," several ISO 9001:2015 clauses lack clarity in defining requirements: the annex to ISO 9001:2015 allows avoidance of the requirement for risk-based thinking, ISO 901:2015 does not contain "common core definitions," and prescriptive "notes" are found in ISO 9001:2015 that do not conform to Annex SL guidelines. The ISO MSSs are required to be updated every five years. The current ISO 9001 standard was approved as an American National Standard on October 21, 2015, and is due for reissue in 2020. TC 176 should start working on the next revision quickly. Several chapters in *The Rise and Impending Demise of ISO 9001* have clearly pointed out the defects in ISO 9001:2015.

1 ISO.org.

ISO 9001 should be reissued using the ISO 9001:2008 standard as a baseline:

- Risk analysis, combined with change control, should replace the 2008 requirement of preventive action, using the ISO 14001:2015 "Actions to Address Risks and Opportunities" as a template, with the clarifications noted in this book.
- The organizational knowledge requirement of ISO 9001:2015 should be included.
- The prescriptive "notes" of ISO 9001:2015 should be removed.
- The ISO 9001:2008 requirements of procedures, records, and quality manual should replace the ISO 9001:2015 terminology of "documented information."
- The 2015 requirement to plan how to achieve its quality objectives should be included.

Oversight of US TAG 176

In the US, ANSI provides oversight to the US TAGs, including US TAG 176. Through ANSI, the US has immediate access to the ISO and IEC standards-development processes. ANSI participates in almost the entire technical program of both the ISO and the IEC and administers many key committees and subgroups. Part of its responsibilities as the US member body to the ISO include accrediting US TAGs, whose primary purpose is to develop and transmit, via ANSI, US positions on activities and ballots of the international technical committee.

The responsibilities described for ANSI would indicate, since ANSI provides *oversight* to the US TAGs, that they are the TAG's supervisor and are responsible for any products that are prepared or approved by the US TAGs, particularly since ANSI sells the standards approved by the US TAGs. I am familiar with other standards administered by ANSI, and I believe the standards released by ANSI were consistently of professional quality. ISO 9001:2015, as described in this book, has many errors, formatting issues, contradictions, and convoluted requirements.

After several communications with the administrator of ANSI in which I described my concerns with ISO 9001:2015, it became clear that ANSI administers the *process* for establishing the US TAGs' position on the preparation of the ISO management standards but takes no *ownership* of the content of the product. US TAG 176 was solely responsible for the United States' position on both the content and approval of ISO 9001:2015. See Figure 15.1 for an email correspondence from ANSI regarding ANSI oversight of US TAGs.

I recommend that ANSI expand the scope of its oversight of the US TAGs to include a quality/technical review and approval of the standards issued by the US TAGs and sold by ANSI.

From: [Name redacted] @ansi.org
To: dentchm
Sent: Fri, Aug 10, 2018 5:14 pm
Subject: FW: Request for clarification of ANSI's Oversight of the US TAGs

Dear Mr. Dentch,

I'm writing in response to your communication below and follow-up letter to President Bhatia. Apologies for the delay, as your request was misdirected.

By way of background, ANSI is the US Member body to ISO, but ANSI does not write the technical content of ISO documents or formulate technical positions in relation to ISO documents. Nor does ANSI, as an Institute, approve such things as "notes" in ISO standards. US positions on proposed ISO standards are formulated by ANSI-Accredited US Technical Advisory Groups (TAGs) that are administered by US TAG Administrators, i.e., organizations such as ASQ, who serve in that role as defined in ANSI's International Procedures (www.ansi.org/internationalprocedures). A list of all ANSI-Accredited US TAGs along with a contact for each is available at www.ansi.org/isotags. It is these US TAGs that determine the final US position and comments submitted to ISO, with respect to any proposed ISO standard. ANSI transmits that vote and any comments to ISO as the US National Body to ISO, but our role is not to formulate technical positions or votes. With respect to sales, ISO National Member Bodies are authorized to sell ISO documents. ANSI does not choose which documents it sells based on a technical assessment of the contents.

As it appears that you are already familiar with ASQ and the standards writing process as well as other US TAGs, I suggest that you submit your proposed revisions to the documents, to the appropriate US TAG Administrator directly and in the required format, for consideration by the respective ANSI-Accredited US TAGs. To understand the applicable procedures, please review the ANSI International Procedures and also contact the TAG Administrator for any additional procedures, forms and deadlines that may apply.

For more information about ANSI, please review www.ansi.org/WhatIsANSI? I would be glad to answer any further questions, so please send your questions to me directly.

I hope that this information is helpful.

Sincerely

[Name redacted]

Figure 15.1: Email Correspondence from ANSI on ANSI Oversight of US TAGs

The ISO Survey of Management System Standard Certifications

Chapter 6, "The ISO Survey of Management System Standard Certifications (2018)," describes the inaccuracies in the data furnished by the CBs related to the number of organizations that hold ISO 9001 certificates. The CBs have confounded the data over several years by reporting both the number of companies that have certificates and the number of sites (plants) that are certified. Some CBs do not fully report the number of certificates, whether a site or a company.

A key component of a CB's guidance for the issuing of a certificate is its verification of the scope of the company's processes, with industry-code classification confirmed to match the scope. The industry sector provides the CB with evidence that the audit days assigned are consistent with the auditee's manufacturing or service activities—and the assigned auditor's qualifications. The 2018 ISO survey data reports that considerable ISO 9001 certificates have been issued in several countries without the identification of industry-sector codes.

The lax data collection and reporting of the ISO survey is disappointing. To report that 10 percent of companies currently certified to ISO 9001 are holding bogus certifications should be unacceptable to ISO leadership. In some countries, the bogus certificates represent more than half the total of their certified companies. The CBs I have worked with were quite insistent on ensuring that the scope of activities and industry sector for organizations they would certify were accurately identified.

The accreditation bodies (ABs) should hold the CBs accountable for accurate reporting of the number of organizations they certify to ISO 9001, and they should ensure that the scope of activities of the certified organizations are accurately defined.

The financial model for the ISO process starts with the company being audited paying the CB, which pays the third-party auditor; the AB receives payment from the CB, based on the number of clients certified by the CB. I have been witnessed by ABs during audits. If I made mistakes, or misapplied the ISO 9001 requirements during the audit, then the AB would issue nonconformances against the CB that had hired me. If there were systemic issues with other aspects of this CB, then the AB could suspend or drop accreditation of this CB, resulting in financial and marketing damage to the CB. When an AB audits a CB, the data discovery should include the CB's business performance: How many clients were added or dropped from the previous year's total? If the CB's client count decreased, then what are the causes—and corrective actions?

The AB auditor could sample several newly issued certificates and audit reports and match the certificate scope against the audit report. If the defective certificates were anywhere near 10 percent of the total, the AB auditor would easily locate the errors and issue nonconformances. If the ABs aren't successful in quickly improving the quality of the ISO Survey of Management System Standard Certifications, then the International Accreditation Forum (IAF) should hold the ABs accountable.

Man-Days Required for ISO 9001 Third-Party Audits

In my survey of ISO 9001 end users, 50 percent of the participants cited that the cost of third-party audits were not justified by the perceived value to their business gained by obtaining ISO 9001 certification. If their customers would not require ISO 9001 certification, then 36 percent of the respondents indicated that they would drop their ISO certificate.

When planning an audit, the CB (or registrar) uses the auditee's employee headcount as the starting point for determining how many days the auditor(s) will be required to spend at the auditee's site (refer to Table 9.1 in chapter 9). Examples of situations that may be used to justify less time include:

- The organization does not design products
- Processes involve a single function activity, such as service only
- The management system is mature
- A high percentage of employees do the same tasks

The maximum reduction from the man-day is 30 percent, and the CBs need to be careful to document the justification for the reduction, as experienced AB auditors will look for issues where CBs might be overly reducing man-days to gain a cost advantage with the competition. There are also situations where the CBs will add audit times above the time specified:

- more than one building or remote design location
- a multilanguage worker base
- logistics considerations with large complexes such as chemical plants
- high degree of regulation (food and drugs, aerospace, nuclear power, etc.)
- high number of unique activities/processes
- remote personnel (home offices technical service)

In my experience with various CBs, they are more prone to add audit days, particularly when the auditee has remote locations, than agree to a reduction due to simplicity or redundancy of the auditee's processes. I was often criticized by my CB supervisor for lobbying on behalf of the audited organization to reduce audit days. My motivation was not so much for saving the auditee money, but rather, I was often embarrassed to attempt to generate productive audit time. As a contractor, I would turn down auditing assignments where I would be obliged to fill up three audit days when I knew I could provide an effective assessment in two days or less. Sitting alone in the auditee's conference room, reviewing dozens of documents, is not a pleasant way to spend a few hours. I often audited with a colleague at large companies where many employees would be doing the same tasks. My colleague would spend a few hours interviewing almost every employee in the company, quizzing them on the company's quality policy. Sampling a few employees in each department provides evidence of the effectiveness of the company's communication process, so my colleague was just artificially creating auditing time without adding value to the auditee's QMS.

Current Process to Establish Audit Days

Currently, the process used to establish audit days for ISO 9001 audits is centered on the number of employees in the audited company. While adjustments are possible, the auditee's headcount dominates the calculation. When I was conducting several third-party audits a month, I developed a protocol for creating audit plans. I would be assigned to audit a very small machine shop for one day, and then several days at a large semiconductor factory. My audit plan for organizations of all sizes would be quite similar for several of the company's processes. My day would start with an opening meeting, a site tour, and an audit of the company's management responsibilities. This portion of my audit would be complete by lunch. I would then start auditing the various processes connected to the customer, including sales, design, and purchasing. Each core process would require between one to two hours for an experienced auditor to assess the auditee's conformance to the ISO 9001 clause requirement. Additionally, each support process of documentation, corrective action, internal auditing, training, maintenance, and calibration can be audited adequately in about an hour.

What remained to be audited was the organization's manufacturing processes. In a single-process organization, such as a machine shop, the manufacturing audit would require one to two hours of audit time. If the company was a multiprocess shop—for example, an original equipment manufacturer (OEM)—then the manufacturing areas would require several hours to sample, including machining, welding, and assembly. The audit of purchasing for the OEM would also require more audit time than the machine shop due to the variety of materials used by the OEM.

The days required for auditing the design process creates another challenge for auditors. Take a plastics-molding company with fifty employees as an example. From Table 9.1 (in chapter 9), the audit days would be five days on-site if the organization has fifty employees and provides designs for the products it molds. If the organization with fifty employees molded products using its customers' designs, then the audit days would be reduced by 20 percent, or one day. An auditor would be hard-pressed to find eight hours of useful design-audit activities in the plastics-molding company. The design processes for plastic molders or similar shops usually involve minor changes to existing products—certainly not as complex as a high-tech company that produces sophisticated semiconductor materials.

Audit Man-Day Planning

The IAF procedure (IAF MD 5:2019) allows the CBs to establish audit days based on the risk category of the organizations processes. According to 1.10, "Risk Category (QMS Only)":

> For QMS, the provisions in this document are based on three categories, dependent on the risks posed by failure of the product or service of the client organization. These categories can be considered as high, medium or low risk. High risk activities (e.g., nuclear, medical, pharmaceutical, food, construction) normally require more audit time. Medium risk activities (e.g., simple manufacturing) are likely to require the average time to carry out an effective audit and low risk activities less time. (See Annex A, Table QMS 2.)

Procedure IAF MD 5:2019, Annex A, Table QMS 2, subjectively defines three risk levels (high, medium, and low) without providing guidance on how to use the risk levels to adjust man-days: "These risk categories are not definitive; they are examples only that could be used by a CB when determining the risk category of an audit." The motivation for the CBs to apply the risk screening to reduce the audit days is low. The audit-day charges related to audit-day planning are what drive the financial engine for the entire ISO pipeline: the CB, the CB's auditor, *plus* the fees charged by the ABs, which ensure that the CBs are providing proper audits.

The process used to establish ISO 9001 audit-day requirements should more accurately reflect the audited company's processes and quantified risk levels.

Environmental and OH&S Audit-Day Planning

When I conducted ISO 14001 and OHSAS 18001 audits, the audit days assigned to me usually resulted in a full schedule, as the environmental and OH&S audit-planning process defined the audited client's complexity of processes: high, medium, low, or limited (Table: 15.2). The CB would review the organization's scope of activities against the Standard Industrial Classification of economic activities to determine complexity. The SIC code translates to the economic activity (EA) utilized in most countries. For example, a manufacturer of basic chemicals, pesticides, and paints would fall under EA 12 and be designated as high risk; a manufacturer of rubber or plastics products would be designated EA code 14 and low risk. The chemical manufacturer with fifty employees would require eight audit days for the initial certification audit, the rubber manufacturer 4.5 days. The difference of 3.5 days for the chemical manufacturer would be used to audit the operating controls related to possible hazardous waste, air controls, chemical containment, and waste-water controls, as well as compliance regulations. The EMS audit-day process provides a quantifiable rationale for establishing audit time. The audit-day-calculation process used by the QMS is subjective and open to inconsistent application, as well as bias toward establishing audit days at the high end to increase revenue.

Annex B: Environmental Management Systems Table EMS 1: Relationship between Effective Number of Personnel, Complexity, and Audit Time (Initial audit only: Stage 1 + Stage 2)				
Effective No. of Personnel	**Audit Time: Stage 1 + Stage 2 (days)**			
	High	Med.	Low	Lim.
1-5	3	2.5	2.5	2.5
6-10	3.5	3	3	3
11-15	4.5	3.5	3	3
16-25	5.5	4.5	3.5	3
26-45	7	5.5	4	3
46-65	8	6	4.5	3.5
66-85	9	7	5	3.5
86-125	11	8	5.5	4
126-175	12	9	6	4.5
176-275	13	10	7	5
276-425	15	11	8	5.5
426-625	16	12	9	6
626-875	17	13	10	6.5
876-1,175	19	15	11	7
1,176-1,550	20	16	12	7.5
1,551-2,025	21	17	12	8
2,026-2,675	23	18	13	8.5
2,676-3,450	25	19	14	9
3,451-4,350	27	20	15	10
4,351-5,450	28	21	16	11
5,451-6,800	30	23	17	12
6,801-8,500	32	25	19	13
8,501-10,700	34	27	20	14
>10,700	Follow progression above			

Table 15.2: ISO 14001:2015 Audit Days Man-Days (IAF MD 5:2019)

Epilogue

I have a difficult time envisioning a scenario where ISO 9001 certifications, particularly in the US, can attract new users or regain the twenty thousand or so companies that have dropped their certificates over the last many years. At best, the ISO process can be improved (as I have suggested in this book) to stem the tide of companies discontinuing ISO 9001 certification. Possibly, service-related companies in the US can find value in ISO 9001 certification, as is the case in many other parts of the world. The sector-specific standards (TS, AS, and 13485) maintain ISO 9001 requirements as a baseline, but many organizations are convincing their nonsector customers that they don't require an ISO 9001 stand-alone certificate; their TS 16949, AS 9100, TL 9000, or 13485 conformances will assure all customers that the company provides high-quality products.

One change I believe would help companies decide to retain their ISO 9001 certification would be to limit the requirement for most companies to be subject to the third-year recertification system that's currently in place, at a cost of about twice the normal surveillance-audit cost. An exemption process could be established based on the organization's successful performance during the previous two years of surveillance. Since this change would immediately reduce the sales revenue of the CBs, I realize it has little chance of acceptance. The CBs would not accept a few days' loss of revenue per each recertification audit. Unless a dramatic change occurs, the number of ISO 9001-certified companies will continue to decline in the United States—and will continue to decrease on a worldwide basis as well.

The ISO Processes Lack Customer Focus

As I researched and developed the chapters of *The Rise and Impending Demise of ISO 9001*, I became increasingly aware of the irony and hypocrisy of the ISO 9001 process. An organization with a well-managed QMS understands its customers and its requirements, and makes adjustments when the requirements are not being met. The organization will clearly communicate requirements to its internal and external suppliers. The ISO 9001 management chain exhibits several flaws that are inconsistent with a well-managed QMS. The "employees" of the ISO 9001 process—the third-party auditors who execute the requirements of the ISO 9001 MSS—are trained to critique (i.e., by writing nonconformances) when organizations they assess do not know their customers, including their activities or the products they deliver. Unclear instructions to an audited organization's workers commonly result in the issuing of nonconformances.

The Future of ISO 9001

The Rise and Impending Demise of ISO 9001 records my many years of witnessing how ISO 9001, when properly administered, was a useful support tool to an organization's quality management system. When the authors of ISO 9001 attempted to make ISO 9001 the key business driver of the organization and diminished the discipline of clearly defined and documented requirements, ISO 9001 lost its key value. Additionally, ISO 9001:2015 became a risk-centric standard of an organization's business as opposed to the traditional quality management system. The ISO 9001:2015 standard does not adequately define the requirements for assessing risks in an organization's business, and third-party auditors are generally not competent in assessing a company's risks.

As the number of ISO 9001 certifications continues to decline, many companies will still want their suppliers to maintain some form of verification—or certification of the supplier's product or service quality management system. Who or what process will fill the void once ISO 9001 certification audits disappear? There are three options available for customers to monitor supplier quality other than ISO 9001 certification: (1) the customer audits the supplier, (2) the customer employs an organization to audit the supplier, (3) the supplier audits itself.

Customer Audits: Many large manufacturers, particularly companies engaged in production of medical products or automotive or aerospace equipment, maintain an auditing staff that audits their suppliers at some frequency, reviewing the specific quality requirements related to the products made for them. These audits are often conducted at suppliers that maintain ISO 9001 third-party certification. When a supplier to the sector-specific company discontinues third-party certification to ISO 9001, the company will probably increase the frequency of the company-managed supplier audits. Other nonsector companies that rely on the ISO 9001 certification process to validate their suppliers' quality will need to decide whether to initiate supplier audits, or perhaps, increase the intensity of their inspection of the supplier's products. Both options will add cost to the customer's operations, including some expensive travel costs to domestic or international locations.

Supplier Audits: Often referred to as "supply chain audits," the supplier audits are a form of outsourcing audits to another organization by the customers described in the first option. In many cases, the supply chain audits are conducted at very small suppliers (less than ten employees) that don't maintain ISO 9001 certification. I have participated in supply chain audits for a large medical products manufacturer in the Boston area. The audits were one-day audits utilizing the medical company's checklist and report forms. The checklists were narrowly focused on the manufacturing and inspection processes employed at the supplier. The checklists did not contain the ISO 9001 quality management system requirements—what I refer to as "ISO boilerplate" (management reviews, internal audits, customer feedback, risk management). The checklists, audit planning, and audit reports were efficiently designed by the auditing firm that hired me. The supplier audits were a more relaxed and productive experience than many of the third-party audits I have recently conducted for certification bodies. The audit documentation I used at these audits was more informative than the bureaucratic audit paperwork currently employed by most CBs. The suppliers were about twenty-five to fifty miles from my home, so travel expenses were minimal. The medical products company paid the auditing firm that hired me approximately $1,200 for the one-day audits.

As ISO 9001 certifications continue to decline, I believe the supply chain audits could fill the role ISO 9001 certification has played in monitoring the quality of products delivered by small suppliers to the large sector-specific customers (or any large ISO 9001-certified company). The firms that provide the supplier audits need to establish a cadre of experienced contract quality auditors located in relatively close proximity to the firm's audited clients. The nonsector-specific companies that use ISO 9001 certification to validate the quality of their suppliers will most likely conduct the previously described customer audits for the suppliers deemed critical to their finished product quality. For the remaining suppliers, these companies may have the suppliers provide evidence of a self-audit.

Self-Audits: Customers of all types have been allowing suppliers to provide a self- assessment of the supplier's quality system for many years. Typically, when a company is qualifying a new supplier, a questionnaire will be sent to the supplier requesting information on the number of employees in the company, their manufacturing processes, material sources, and details of their quality assurance program. If the supplier is ISO 9001 certified, the customer will request a copy of the supplier's ISO 9001 certificate—and the customer will allow the supplier to skip the questions related to quality assurance. The self-audit questionnaires I have observed are mostly generic, and do not provide much information to ensure the supplier will deliver products of consistently acceptable quality. Companies that elect to drop their ISO 9001 certification should consider auditing their processes at some frequency to evaluate the consistency by which they deliver products to their customers.

Quality System Self-audit: Based on my experiences conducting 3rd party audits, internal audits, and supplier audits of companies with a wide range of manufacturing activities, I created a set of quality system requirements. The requirements and self-audit checklists can be used by organizations to provide an audit of the organization's capability to fulfil customer's purchase orders. I did not include the product design process as the majority of the organizations that may wish to use the self-audit manufacture products based on their customers' designs. The quality system requirements are focused on organizations in the manufacturing sectors. With some adjustments, an organization providing "services" (distributers, engineering firms, etc.) can adjust the checklists to align with the processes central to servicing. "The Quality System Requirements—2020" is outlined below. The file and checklists can be accessed on the link:

http://users.neo.registeredsite.com/9/2/0/18247029/ /1003959/quality-management-system.html?_=0.915278594051067.

Readers of *The Rise and Impending Demise of ISO 9001* with questions or comments can contact me at miltdentch@gmail.com. My website (MPD-QE-Consulting.com) contains many of the guidance documents, procedures and templates referenced in the book.

Quality System Requirements—2020

1.0 Order Entry Process:
Requirement: *The organization shall establish contracts with customers to define the terms for the customer to purchase products from the organization.*

a) the organization shall review each contract (purchase order or tender offer) from customers to ensure that the customer's requirements are adequately defined and documented.
b) the organization shall verify the organization's processes have the capability to meet the customer's contractual requirements.
c) if the customer requirements are changed during the processing of the order, the contract shall be modified to conform to the revised requirements.

2.0 Production Planning Process:
Requirement: *The organization shall plan the processes required to produce the contracted products to ensure that the processes are carried out under controlled conditions to include:*

a) work instructions, defining the method(s) of production.
b) qualified processes, production equipment, and employees.
c) a suitable working environment.
d) the applicable quality standards, regulations, and specifications.

3.0 Procurement Process:
Requirement: *The organization shall ensure materials, products, and services procured to fulfil the customer's contract meet specified requirements.*

a) the organization shall ensure that purchased products, materials, or services conform to specified requirements.
b) the organization shall select and approve suppliers based on the supplier's ability to meet the purchase order requirements.
c) the organization shall monitor the approved suppliers' ability to consistently provide material and services that conform to the organization's requirements.

4.0 Incoming Material Inspection Process:
Requirement: *The organization shall ensure procured materials conform to the organization's specifications.*

a) the organization shall ensure that incoming material from suppliers is not used or processed until it has been inspected or otherwise verified as conforming to specified requirements.

b) the organization shall ensure incoming material is protected and preserved prior to processing.

5.0 Production Processes:

Requirement: *The organization shall implement the manufacturing processes to fulfil the customer's contract by establishing appropriate controls that ensure:*

a) the production equipment and supporting production processes are capable of manufacturing the customer's products.

b) the customer's product parameters, including the physical and workmanship specifications, are monitored and maintained during the manufacturing processes.

c) the product and processed materials are protected during the manufacturing, storage,

d) handling, packaging, and shipping stages prior to, and during delivery to the customer.

6.0 The Process to Establish and Maintain Employee Competence:

Requirement: *The organization shall ensure employees have the skills and competence required to accomplish the assigned tasks to include:*

a) determining the competence for employees processing orders, producing and inspecting products, maintaining the processes and equipment and administering the quality system.

b) determining training needs.

c) updating employee training when procedures or work instructions change.

7.0 Product Inspection Process:

Requirement: *The organization shall verify that product and materials meet the customer and organization's requirements during all stages of manufacturing.*

a) the organization shall inspect, test and identify the product during the processing of the product as required by the instructions for manufacturing the customer's product.

b) the organization shall implement the final inspection and testing processes established to ensure that the finished product satisfies the customer's specifications and the organization's workmanship criteria.

c) the organization shall not release product for shipment to the customer until the organization has completed and documented the final inspection of the product.
d) the organization shall ensure that product that does not conform to specified requirements is prevented from inadvertent use.

8.0 Equipment Maintenance and Device Calibration Processes:
Requirement: *The organization shall ensure production equipment and measuring devices perform as planned.*
a) the organization shall maintain production, plant support equipment to ensure the equipment maintains the process capability required to meet the specifications of the products and materials manufactured.
b) the inspection, measuring, and test devices used to measure and record the parameters, dimensions, or characteristics of finished product and materials used in the manufacturing processes shall be controlled, calibrated, or validated as required to provide evidence of product conformance to customer specifications.

9.0 Document and Record Control Processes:
Requirement: *The organization shall ensure employee instructions are available at point of use, and the instructions accurately describe these procedures and tasks; the organization shall maintain records of the process outputs and measurement and testing results.*
a) the organization shall prepare and control procedures (of internal and external origin) and instructions required to ensure employees have the information required to perform the tasks related to their job responsibilities.
b) the organization shall maintain the records required to demonstrate conformance to specifications and requirements for the input and output of the organization's processes to include order entry, purchasing, production, inspection, training, maintenance and management responsibilities.

10.0 Management Processes:
Requirement: *The organization shall control, review, and support improvements to the organization's performance by:*
a) defining the responsibilities, authority, and the interrelation of all employees who manage, perform, and verify work related to producing customer products.
b) investigating the cause(s) of nonconforming product and processes,

and implement the corrective action needed to prevent recurrence of the nonconformance.

c) monitoring the organization's processes at a defined interval to evaluate the effectiveness of the organization in meeting the organization's customer requirements and the achievement of the organization's performance metrics.
d) establishing and monitoring the metrics and (goals) required to evaluate the performance of the organization's processes.
e) reviewing the performance of the organization's processes to establish a baseline for process changes and improvements.

About the Author

Milt Dentch was born and raised in the Worcester, Massachusetts, area, graduating from St. John's High School. He received a BS in mechanical engineering from Worcester Polytechnic Institute and an MS in quality management systems from the National Graduate School of Quality Management (NGS). After college, he worked as an engineer in the paper industry for several years. Milt was employed by the Polaroid Corporation in Waltham, Massachusetts for 27 years, starting as a plant engineer and retiring as the Director of Worldwide Materials in 1996. After Polaroid, he was plant manager for the Custom Coating and Laminating plant in Worcester for the Furon Corporation for several years.

Milt has over 50 years' manufacturing experience in a wide variety of industries, including pulp and paper, chemical, plastic and rubber processing, battery manufacturing, converting, electronics assembly, and machine building. Milt currently provides consulting, training, and auditing related to the International Organization for Standardization (ISO) requirements for quality, environmental and Occupational Health and Safety (OH&S) management Systems. He was an Exemplar Global (RABQSA) qualified lead auditor for eighteen years, approved to conduct quality, environmental and OH&S management system audits.

Milt has conducted over 500 audits in companies as diverse as a floating oil rig in the Gulf of Mexico and an electronics manufacturer in Ukraine with 4,000 employees.

In 2012, Milt wrote *Fall of an Icon—Polaroid after Edwin H. Land* (RiverHaven Books), an insider's history of the Polaroid Corporation. His books *The ISO 14001:2015 Implementation Handbook*, *The ISO 9001:2015 Implementation Handbook* and *The ISO 45001:2018 Implementation Handbook* were published by Quality Press of ASQ in 2016 and 2018.

Made in United States
Orlando, FL
04 November 2026

90201275R00098